Milnes Marshall

Studies from the Biological Laboratories of the Owens College

Vol. II

Milnes Marshall

Studies from the Biological Laboratories of the Owens College
Vol. II

ISBN/EAN: 9783337216306

Printed in Europe, USA, Canada, Australia, Japan

Cover: Foto ©berggeist007 / pixelio.de

More available books at **www.hansebooks.com**

STUDIES

FROM THE

BIOLOGICAL LABORATORIES

OF

THE OWENS COLLEGE.

VOLUME II.

PUBLISHED BY THE COUNCIL OF THE COLLEGE

AND EDITED BY

PROFESSOR MILNES MARSHALL.

MANCHESTER:
J. E. CORNISH.
1890.

PRICE TEN SHILLINGS.

CONTENTS.

	PAGE

Mr. G. HERBERT FOWLER.—"The Anatomy of the Madreporaria." Part II. Plate I. *Reprinted from the " Quarterly Journal of Microscopical Science,"* 1886 1

Mr. G. HERBERT FOWLER.—"The Anatomy of the Madreporaria." Part III. Plates II. and III. *Reprinted from the " Quarterly Journal of Microscopical Science,"* 1887 17

Mr. ARTHUR ROBINSON.—"On the Position and Peritoneal Relations of the Mammalian Ovary." Plate IV. *Reprinted from the " Journal of Anatomy and Physiology,"* 1887 35

Dr. C. HERBERT HURST.—"The Pupal Stage of Culex." Plate V. 47

Mr. C. F. MARSHALL.—"Observations on the Structure and Distribution of Striped and Unstriped Muscle in the Animal Kingdom, and a Theory of Muscular Contraction." Plate VI. *Reprinted from the " Quarterly Journal of Microscopical Science,"* 1887 73

Dr. A. M. PATERSON.—"On the Fate of the Muscle-plate, and the Development of the Spinal Nerves and Limb Plexuses in Birds and Mammals." Plates VII. and VIII. *Reprinted from the " Quarterly Journal of Microscopical Science,"* 1887 103

Mr. A. E. GILES.—"Development of the Fat-bodies in the Frog: a Contribution to the History of the Pronephros." Plate IX. *Reprinted from the " Quarterly Journal of Microscopical Science,"* 1888 123

Professor MARSHALL and Mr. EDWARD J. BLES.—"The Development of the Kidneys and Fat-bodies in the Frog." Plate X. 133

Mr. FRANCIS VILLY.—"The Development of the Ear and Accessory Organs in the Common Frog." Plates XI. and XII. *Reprinted from the " Quarterly Journal of Microscopical Science,"* 1890 159

Professor MARSHALL and Mr. EDWARD J. BLES.—"The Development of the Blood Vessels in the Frog." Plates XIII., XIV., and XV. 185

PREFACE.

THE majority of the papers in the present volume are reprinted from the *Quarterly Journal of Microscopical Science*, or from the *Journal of Anatomy and Physiology;* and to the editors of these journals grateful acknowledgments are due for permission to republish the papers and the plates illustrating them.

Dr. Hurst's Inaugural Dissertation, and the papers by Mr. Bles and myself, have not been previously published.

Dr. Fowler's contributions on the Anatomy of the Madreporaria represent work done during his tenure of a Berkeley Research Fellowship, a generous benefaction which has already in his own and in other cases yielded excellent results, and which promises to rank among the most successful and most characteristic of Owens College institutions.

The series of papers on the Development of the Frog, perhaps the most conspicuous feature of the volume, contain the results of investigations on which several of my pupils have, with myself, been engaged for some time past.

A. M. M.

Owens College,
    June, 1890.

THE ANATOMY OF THE MADREPORARIA.

II.*

By G. HERBERT FOWLER, B.A., *Keble Coll. (Oxon), Berkeley Fellow of the Owens College, Manchester.*

[WITH PLATE I.]

IN a previous paper (4), I have described the anatomy of a solitary Imperforate coral, Flabellum; and of a branching Perforate, Rhodopsammia. The present memoir treats of two examples of colonial Perforate forms, *Madrepora Durvillei* and *M. aspera.*

MADREPORA DURVILLEI (Milne-Edw. and Haime).

Two fragments of this perforate Madreporarian were kindly entrusted to me for study by Professor H. N. Moseley, who had obtained them during the voyage of H.M.S. "Challenger."

The species was founded by Milne-Edwards (1) from a part of the *M. rosea* of Esper, but as his account is very incomplete, Mr. J. J. Quelch, of the British Museum, has furnished the following description of the coral. I am glad to be able to take occasion to thank him for this and many other courtesies.

A. "CORALLUM: Arborescent, spreading, and remotely ramose, or occasionally sub-prostrate, and almost destitute of branchlets on the under surface. *Branches* often nearly 2 cm. thick, becoming very thin towards their extremity, sub-terete, elongated, covered irregularly with crowded capillary polyp-bearing branchlets, which generally give to the branches a sub-cylindrical outline of about

---

* For Part I. see "Quarterly Journal of Microscopical Science," August, 1885, and "Studies from the Biological Laboratories of the Owens College," Vol. I., 1886.

3-5 mm. in diameter. *Branchlets* small and short, about 1-2 cm. in length, consisting generally of a few thin and long tubiform calicles; towards the apical parts of the branches they become much less elongated and often quite short. *Surface* slightly porous, very distinctly costulated throughout, and marked with fine echinulations which are very distinctly arranged on the calicles. *Calicles* generally tubiform, about 1·5 mm. wide and 1 cm. long, except towards the apical parts of the branches, where they are shorter and smaller, and sometimes tubonariform; a few short tubonariform calicles are generally placed on the surface of the branches between the branchlets. *Star* distinct, of six more or less lamelli-spiniform septa, two of which, the distal and the proximal, are usually much enlarged, and meet one another, often deep down in the fossa; while occasionally, as in the terminal calicles, the six septa are sub-equal, and coalesce at the centre."

"This species seems to be distinguishable from *M. echinata* (Dana) simply by the costulations of the surface, which in the latter is smooth or finely granulated. It is doubtful, however, whether this character will prove to be sufficiently constant to separate the two species, when a larger number of forms has been examined."

Figs. 1 and 2 represent the dorsal and ventral aspects of a fragment of a branch, and show most of the characteristics mentioned in the above description.

In a transverse section of the corallum (Fig. 3), the peripheral ring of polyp cavities is cut somewhat obliquely ($a\ a$), owing to the inclination of branchlets and calicles to the branch; while the more central ones, cut at a lower level and more transversely, are approximately circular in outline ($a'\ a'$). They lie, roughly speaking, on three sides of the branch, none are apparent on the fourth. The shorter radius of the latter seems to imply that the growth in diameter of the branch depends upon the outward growth of the polyps.

In the axis of the branch is a *central cavity* (*c.c.*), into which project six septum-like ridges; this probably represents a cavity previously inhabited by the now apical polyp. The tops of both my specimens having been broken off, I have not been able to prove this; nor again to investigate the method of budding; but in *M. aspera* is such another central cavity with six septa, which is continuous with that of the apical polyp. All other polyp cavities converge towards, and, by means of canals, eventually open into, this central cavity, but no more definite connection is traceable. Tissues not unlike mesenteries are sometimes visible in it, but the alcohol in which the specimens were killed did not penetrate sufficiently rapidly to preserve the central parts in good histological condition. In some sections the six septa are not recognisable, and the axis of the branch is occupied by a wide-meshed network of coral; this is probably due to re-absorption of part of the skeleton.

In transverse section are also seen concentric series of *longitudinal canals* (c^2) permeating the corallum; their arrangement appears to indicate that the radial growth of the branch is effected in the following manner. Directly beneath the external body wall of the colony a series of longitudinal canals runs between the costæ (Fig. 4, c^1); and it is probable that, for increase in the diameter of the branch, the costæ grow outwards, and then, bulging laterally, fuse over these canals, so as to enclose them entirely in corallum (*cf.* Fig. 10, *x*). Thus there results a series of internal longitudinal canals, concentrically arranged, with radii of coral between them which represent former costæ. Not only does the appearance of such a transverse section as Fig. 3 suggest that this is the mode of growth, but also "dark lines of growth" (Fig. 5) run radially from each costa towards the centre, so continuously as to indicate that what was a costa when the diameter of the branch was very small, has continued to grow as such, and to be still such when the diameter is very much larger. New costæ, when required owing to the increased circumference of the branch, appear to take their origin from the point of fusion of previous costæ.

More minutely, growth is effected, presumably by the activity of calycoblast cells, through the addition to and formation of *crystalline ellipsoids*, similar to those described by von Koch in Stylophora (**2**). These ellipsoids have a distinct sweep from one "line of growth" to the next.

The *calyces* are all of approximately the same size, and that so minute as to render investigation of the anatomy difficult.

The *septa* are very irregular of occurrence; the complete number appears to be six, but three are rarely to be seen in one section, often none at all. They are not constant through the whole depth of the polyp cavity, but occur as discontinuous ridges (Fig. 6, *Ab*). In every polyp, however, either an axial or abaxial septum is present, which enables the orientation of the polyp to be effected as in the Alcyonaria. (These terms, axial and abaxial, are used in preference to the ordinary and misleading "dorsal" and "ventral," and were suggested originally by Professor Milnes Marshall, "Trans. Roy. Soc. Edin.," 1883.)

There is no columella, but often the axial and abaxial septa fuse, low down in the polyp cavity, so as to divide it into two equal halves (Fig. 3, *a'*), in a manner suggestive of the "median plate" in Pocillopora and Seriatopora figured by Professor Moseley (**3**).

The *costæ* bear apparently no relation to the septa in the well-grown colony, whatever may have been the case in the founder-polyp. Not only is no connection traceable between them in a transverse section of the branch, but even in a single polyp standing off from the stem, where the number of septa is under the most favourable conditions but six, about twenty costæ surround the calicle.

B. ANATOMY.—The whole of the corallum is covered externally by a definite *body wall* of ectoderm, mesoderm, and endoderm (Fig. 6, *ext. b. w.*, Fig. 4, *ect. me. en.*), immediately beneath which lie, as in Rhodopsammia, *external longitudinal canals* parallel to the long axis of the corallum (Figs. 3, 4, 6, c^1). These, however, are not the result of the same anatomical relations in both cases; in Rhodopsammia, lamellæ of mesoderm with a layer of endoderm on each side are given off from the external body wall, and unite with the endoderm and mesoderm which clothe the exterior surface of the theca; and into the canals thus formed project the costæ. In *M. Durvillei*, the layer of endoderm and mesoderm which is immediately apposed to the exterior surface of the corallum, rises in a ridge towards the external body wall; and at the points where these layers meet and fuse are formed the costæ, *i.e.*, in the angle of the mesoderm; and therefore between the costæ lie the canals. A comparison of Fig. 4 with (**4**) Fig. 17 will make clear the anatomical difference.

There is thus no trace of any structure resembling the "peripheral continuations of the mesenteries of von Koch."

These canals appear to open over the lip of the calyces into the polyp cavities; they are connected with each other transversely between the spikes (echinulations) of the costæ (Figs. 6, 7); and further, by radial canals (Figs. 3, 4, c^3) they open into the internal longitudinal canals, which I believe, as above stated, to have, at an earlier period in the history of the branch, occupied a position similarly external to the corallum. The whole system which thus perforates the corallum, and allows free current of fluid to even the most remote parts of the colony, is lined by endoderm and mesoderm throughout, and opens into similarly lined polyp cavities.

The general structure of the colony is, therefore, (1) an *external body wall*, under which and between the costæ lies (2) a series of *external longitudinal canals* opening into each other, and also through the corallum, into (3) the *internal canals*, mainly longitudinal, with radial and transverse connections, communicating in their turn with (4) the *cœlentera of the polyps*. Into the last the external longitudinal canals also open directly, through the theca. The whole system is of course merely a complication of the primitive cœlenteron.

Of the *polyps* there are at least two distinct types, which are full of interest as constituting the first record of marked dimorphism among the Madreporaria. Both are Actinian in structure.

Type A has in the highest sections twelve perfectly normal mesenteries, and a stomatodæum which is a simple invagination of the external body wall. A little way down in the polyp, six of the mesenteries, in every case the same six, assume a curious modification of structure, which will be described first as seen in a series of transverse sections. Fig. 8 represents the characteristic features of a polyp of this type; the mesenteries numbered 2, 4, 6, 7, 9, 11 are those which undergo modification, and are diagrams of a series of drawings made from the same mesentery with camera lucida at different heights.

There appears first (Fig. 8 2) an involution of the stomatodæum directed towards the mesentery, on the floor of which the ectodermic cells are long, but shorter at the sides. By fusion of the mesoderm and obliteration of the ectoderm on each side of this involution, a small canal with a definite lumen is found to be pinched off, and to lie enclosed in the mesoderm lamella of the mesentery (Fig. 8 4). In the neighbourhood of this involution, the endodermic cells lining the mesenterial chamber become enormously lengthened and vacuolated, though the layer is still apparently only one cell deep.

Some sections lower down in the polyp (Fig. 8 6), another similar involution appears in the stomatodæum, in which the ectodermic cells are short on the floor, but pass into deeper ones at the sides; this similarly results in the enclosure of what appears to be a second canal in the centre of the mesentery (Fig. 8 7). In the first canal, as is shown in the diagram, the longer ectoderm cells face towards the stomatodæum; in the second away from it.

Further down yet, where the stomatodæum ceases, the free edge of the mesentery is enlarged into a perfectly normal filament (Fig. 8 9); and finally (Fig. 8 11), the whole modification disappears suddenly, the two canals meeting below; the mesentery then presents a perfectly normal appearance, namely, a mesoderm lamella with a layer of small endodermic cubical cells on each side of it, and bearing the usual filament.

The compilation of these sections, which I have attempted to express in Fig. 6 M, shows that on an ordinary mesentery occurs a swelling due to elongation of the endoderm cells, through which runs, in the mesoderm, a canal lined by ectoderm, doubled back on itself, and opening at both ends into the stomatodæum, with the ectoderm of which its lining is continuous.

Of twenty-one polyps examined, seven present this modification of six (and in all cases of the same six) mesenteries, namely, those numbered 2, 4, 6, 7, 9, 11, according to the method employed in the diagram; the other six mesenteries, 1, 3, 5, 8, 10, 12, and all the twelve mesenteries of the other polyps, are perfectly normal, and show no tendency to such a modification. Were it possible to explain the sectional appearances by a contortion of the mesentery, the regularity with which it occurs would be sufficient proof that it is a definite modification of structure, the parallel of which has yet to be sought in the Anthozoa.

The unmodified mesenteries in Type A generally die out before the plane of the opening of stomatodæum into cœlenteron is reached, in transverse sections. If they present a filament, which is seldom the case, it is of the same character as that figured (Fig. 8 11), i.e., identical with that of a modified mesentery; more frequently none is present, or at most a slight endodermal swelling on the free edge.

The mesenteries 4, 9, run very much deeper into the corallum than the others.

Type B, of about the same diameter as A, is of the normal Actinian structure. The twelve mesenteries are simple, and exactly like those unmodified in Type A. Most of them die out after a very short course, but those numbered 2, 4, 6, 7, 9, 11, on the same notation as in Fig. 8, present a more developed filament than the other six, and extend further down into the corallum, and of these 4 and 9 have by far the longest course, and are the only ones that bear ova.

We have thus two distinct types of polyp, the one distinguished only for entire normality; the other with a hitherto undescribed form of mesentery. In both is observable a differentiation affecting the same six mesenteries, exhibited in the one case as a tendency to a longer course, and to the more complete development of the filament; in the other as the peculiar modification described above; and in both types two of these six have of all the longest course, and are, so far as I have observed, the only ones that bear reproductive organs.

Neither type is confined to certain areas of the branch, but both appear to be irregularly distributed.

Tentacles are not recognisable in my specimens, but it is probable that in the living animal they occur as slight evaginations of the chambers, and have shrunk under the action of the alcohol in which the polyps were killed.

Muscles are obviously present on the mesoderm lamella of the mesenteries, but owing to their minute size it is impossible to detect how they are arranged. I see no reason to doubt that they agree with Actinia. So far as it is possible to judge without this clue, the septa are entocœlic.

c. HISTOLOGY.—There is but little to be said under this head, except as regards the modified mesentery, an almost transverse section of which is represented in Fig. 9. The state of the specimens did not allow of an exhaustive study of cell structure, but those cells, the elongation of which causes the peculiar swelling on both surfaces of the mesentery, are apparently simply lengthened, much vacuolated and amœboid at their free ends. No food particles were detected in them, or indeed in any other part, but many zooxanthellæ are embedded amongst them. These cells pass gradually into the ordinary endoderm, and their appearance suggests strongly that their condition is merely an exaggeration of that of the "Flimmerstreifen" of the brothers Hertwig, *i.e.*, of the two lateral lobes of the mesenterial filament.

In a recent paper (5) Dr. Wilson has suggested that these lateral lobes are ectodermic in origin, circulatory in function, and homologous with the "ectodermic bands" described by him on the axial mesenteries of certain Alcyonaria. I may here state that, so far as histological evidence from the adult is valuable, it points, in all the Madreporaria that I have yet examined, distinctly in the other direction. The central "Nesseldrüsenstreifen" have precisely the same microscopic appearance as the stomatodæal ectoderm; while the "Flimmerstreifen," in the unbroken gradation by which they pass into the endoderm, and by their characteristic staining, seem to be much more nearly connected with that layer than with the ectoderm, and to exhibit an intermediate condition between the ordinary cubical or pavement cells of the endoderm and the enormously lengthened cells of *M. Durvillei*. Von Heider (6), on the same grounds, had previously come to the same conclusion with regard to Cerianthus.

The ova, which in my specimens were few in number, are surrounded by a mesodermal capsule, and possess the ordinary structure. In the one case, in which an ovum was observed on a modified mesentery, it was borne on the neck between the endodermic swelling and the mesenterial filament.

D. GENERAL CONCLUSIONS.—This form has four interesting features in common with the Alcyonaria (Octactiniæ) :—

1. The marked tendency to an absence of polyps on one (the ventral) side of the branch and branchlets.

2. The very definite orientation of the polyps by a stronger development of axial and abaxial septa; and the concomitant bilateral symmetry, the plane of bisection being at right angles to the long axis of the branch or branchlet.

3. The differentiation of mesenteries, which, confined in the Alcyonaria to two, is here extended to six, and more particularly to two of these, though not the same two as in the other group.

4. The distinct dimorphism.

Of the true significance of this dimorphism no certain explanation can be gathered from this form studied merely by itself; it can only be resolved by a comparative study of allied species. Differentiation of function appears to be incomplete; both forms are reproductive, both apparently digestive. The most that can be said is that A is, perhaps, more digestive and less reproductive than B, for the filaments are more developed than in the latter form, and I have only once observed an ovum on a modified mesentery. Should the modification be digestive in function, as is probably the case, A might certainly be termed a "gastrozooid."

But at present any explanation of the function of the structure above described, cannot be other than a mere speculation. It cannot be regarded as a necessary result of the colonial habit, since nothing similar occurs in the next species to be described—*M. aspera*. It can hardly be connected with reproduction, as ova are of rarer occurrence in the modified than in the unmodified polyps; and an excretory apparatus is not required by an organism whose cells are capable of amœboid activity, egestion as well ingestion.

The only evidence on the point is derived from the distribution of the zooxanthellæ. These are most plentiful, firstly, in the external canals just under the body wall; and secondly, among the elongated cells of the mesentery. Assuming, as we may fairly do, that nutriment and aeration were the determining factors of such distribution, it would seem that, in the first case, there must be a strong current of nutritive "chyleaqueous fluid" (to use a word of the older zoologists) in these external canals, and that aeration was effected by diffusion of oxygen through the body wall from the surrounding medium; and in the second place, that the elongated vacuolated cells of the mesentery were in some way assimilative, while oxygenation of the tissues for these special digestive processes (and therefore secondarily and accidentally to the benefit of these symbiotic algæ) resulted from a constant stream of water flowing through the central ectodermal canal of the mesentery.

That such a stream does pass through this canal is extremely probable, for the longer ectodermic cells are all morphologically on the same side of the canal; a wave of ciliary action must therefore result in a current through the canal from one of the apertures into the stomatodæum towards the other. A comparison of Fig. 8 7 with Fig. 6 M will explain this arrangement of the cells.

It is interesting to note that in *M. Durvillei*, as in Alcyonaria and Antipatharia, two mesenteries are distinguished from the rest by running far deeper into the corallum or rachis. This may be a specialisation for circulatory purposes, as has been shown by Dr. Wilson to be true for certain Alcyonaria, or connected with production of the generative elements, as in the case in Antipatharia; in *M. Durvillei* certainly the latter, perhaps also the former, holds good.

Madrepora aspera (Dana).

For a fragment of this coral, fortunately the upper part of a branch, I am again indebted to Professor Moseley.

The species was founded by Dana (7), who gives a good figure of the colony.

A. Corallum.—A transverse section of the *corallum* (Fig. 10) shows that the polyp cavities ($a\ a'$) are arranged in a definite ring, and not merely confined to three sides as in *M. Durvillei*, round a central cavity into which project six septa, more or less fused together at their free edges. This *central cavity* (*c.c.*) is continuous with that of the apical polyp of the branch. The arrangement of the internal longitudinal canals is not so definitely concentric as in *M. Durvillei*, but the method of circumferential growth of the corallum appears to be similar in both species, since the costæ appear to fuse over the external longitudinal canals. (*Vide* Fig. 10, x, and p. 3).

In the apical polyps are found six distinct entocœlic *septa*, and six smaller exocœlic, of which all are not always present; in the others generally only an axial or abaxial septum. A similar difference between them was observed by von Koch (8) in *M. variabilis*, where both exosepta and entosepta were present in the apical polyps, but entosepta only in the rest.

In this form, as in the former species, there appears to be no relation in number and position between costæ and septa, the former being by far the most numerous.

The *costæ* are apparently formed as in *M. Durvillei*, that is, at the points where the endoderm and mesoderm apposed to the exterior surface of the corallum touch the external body wall (*vide* p. 4 and Fig. 4), but in both species, owing to alcoholic contraction, the latter has so shrunk on to the corallum that the costæ project through it, and the exact conditions are difficult to determine with certainty.

B. ANATOMY.—The general anatomy of the colony, as regards the relations of canals, body wall, polyp cavities, &c., agrees with that of *M. Durvillei.* Beyond the fact that in *M. aspera* the polyp cavities are placed closer together, and that therefore there are fewer canals in the corallum, there is little or no difference between them. As regards the polyps, however, there is no dimorphism; all the polyps, except those which are obviously immature buds, are identical in structure.

A typical *polyp* possesses twelve perfectly normal mesenteries, and a stomatodæum which is a simple invagination of the external body wall. When numbered on the same system as in *M. Durvillei,* it is found that those mesenteries marked 1, 2, 4, 6, 7, 9, 11, 12, are the ones which develop mesenterial filaments, that is, the same mesenteries as in *M. Durvillei,* with the addition of the abaxial "directives;" while the others, 3, 5, 8, 10, generally have no filament, and do not extend to the bottom of the stomatodæum.

The apical polyps are about twice the size of the others, but, except for their possession of more septa, are identical in structure with them.

The *muscles* in both apical and lateral polyps are arranged on the mesenteries just as in Actinia, and present nothing unusual in structure.

Tentacles I was unable to recognise, macroscopically or by sections, but a figure by Dana shows that they are present, and twelve in number. In this, as in the species last described, they have shrunk into insignificance, owing to the action of the spirit in which the specimens were preserved. They agree with *M. variabilis,* in which, according to von Koch, they are also exocœlic and entocœlic.

The histology calls for no remark, agreeing with that of forms already described. Calycoblasts were very distinctly present in the growing parts of the colony.

C. METHOD OF BUDDING.—With regard to this, I have been able to glean but little information; since the immature polyps are so crowded with zooxanthellæ, owing presumably to the amount of nutriment supplied to them, that the tissues are much obscured.

The stomatodæum is invaginated to a considerable depth into the future polyp cavity before it is perforated for communication between the cœlenteron and the exterior, and also apparently before any mesenteries are formed. The cavity into which it is invaginated is already of considerable diameter, and larger than the ordinary canals of the colony; though smaller than that of a fully formed polyp, at that point it is probably never enlarged by re-absorption of coral, but its continuation upwards by future growth of the polyp possesses a gradually increasing diameter.

In a young polyp in which the stomatodæum was invaginated, but not yet perforated below, the latter appeared to be supported by tissue surrounding the future septa, just as the external body wall is supported by tissue enclosing the costæ. In sections below the stomatodæum, and unconnected with it, were seen two small mesenteries with filaments, which appeared to be growing upwards towards the stomatodæum, and to have not yet joined it. It is therefore possible that these grow upwards from the canal system, and are formed quite independently of the rest of the polyp. This view is further supported by the observation that, in sections quite at the top of a branch, above the plane of any lateral polyps, occur in the canals one, sometimes two, little mesenteries with filaments, which I believe to be growing upwards towards the sites of future polyps. They appear to take rise, near the cavity of the apical polyp, from the wall of the canals.

In the only other stage of development from which any observations could be made, six mesenteries had appeared; of these the two furthest from the axis carried muscles on the outer faces, though it does not necessarily follow that they were the abaxial "directives" of the adult. The muscles of the other two pairs were not sufficiently developed to allow of their arrangement being recognised.

CONCLUSION.—From *M. Durvillei*, the present species is widely separated by a strong morphological distinction, the absence of dimorphism; since the difference between the apical and lateral polyps in *M. aspera* is hardly strong enough to be reckoned as such. That such a distinction should exist between two species of a genus is very remarkable; but, considering the great antiquity of these forms, the similar structure of the colony in both, and the fact that they exhibit a similar differentiation of certain mesenteries, it is not to be inferred that their systematic relations are unsound.

Note.

For microscopic sections through both hard and soft parts of the coral, such as are figured in (4) Pl. XLI., Figs. 14, 15, I have found the method, originally applied by von Koch to these forms, extremely useful. The coral, having been left in borax carmine for three days, and treated with acidulated alcohol for six hours, is transferred to absolute alcohol, and from this to ether; into the ether is dropped *absolutely dry* powdered Canada balsam in small quantities at a time, till enough is dissolved to make a block, rather larger when dry than the specimen. The ether is driven off by a gentle heat, leaving the coral permeated throughout by balsam. About a week should be devoted to this part of the process.

Sections are then cut with a lapidary wheel, or, if this is not procurable, with a fret saw; and ground like geological sections on a slate, then polished on a water of Ayr stone. Oil and emery powder should be avoided, water alone being used for the stones.

One surface of the section having been ground and polished, it should be affixed permanently by that surface to a glass slide, on to which some dry Canada balsam has been melted, and not again be moved. When the other surface has been similarly ground and polished to the required thinness, it should be brushed lightly, first with absolute alcohol, then immediately with oil of cloves; this removes all dirt from the surface. A drop of balsam in benzole is then placed on the section, and the cover glass lightly dropped on it.

Erratum.

In my previous paper (4) Pl. XL., Fig. 1, the septa were wrongly numbered; they should have been marked 1, 4, 3, 4, 2, 4, 3, 4, 1, reckoning on each side from the central "directive" septum, D.

Literature.

1. Milne-Edwards.—"Hist. Nat. d. Coralliaires," iii., p. 148.
2. Von Koch.—"Jen. Zeitschr.," Bd. xi.
3. Moseley.—"Quart. Journ. Micr. Sci.," Oct., 1882.

4. Fowler.—"Quart. Journ. Micr. Sci.," Oct., 1885 ; and "Studies from the Biological Laboratories of the Owens College," Vol. i., 1886.
5. Wilson.—"Mitth. Zool. Sta. Neap.," Bd. v.
6. Von Heider.—"Sitz. k. Akad. Wissench.," 1879.
7. Dana.—"Zoophytes of the Wilkes Expedition."
8. Von Koch.—"Morph. Jahrb.," Bd. vi.

DESCRIPTION OF PLATE I.

a. Polyp cavities, cut obliquely. a'. Polyp cavities, cut transversely. Ab. Abaxial (ventral) septum. Ax. Axial (dorsal) septum. C. Costæ. $c.c$. Central cavity, continuous with the apical polyp. c^1. External longitudinal canals between the costæ. c^2. Internal longitudinal canals. c^3. Radial and transverse connecting canals. Co. Corallum of the main branch. d. Cut edges of the endoderm and mesoderm lining the cœlenteron. ect. Ectoderm. en. Endoderm. $ext. b. w$. External body wall of ectoderm, mesoderm, and endoderm. M. Mesentery, showing the endodermal swelling. me. Mesoderm lamella. S. Septum. $S. C$. Septal Columella-plate. St. Stomatodæum. Th. Theca of polyp. Z. Zooxanthellæ. x. Fusion of costæ over ext. long. canals. All except Fig. 10 are from *Madrepora Durvillei*.

Fig. 1. Dorsal view of the corallum of two fragments of a branch, bearing calicles, and branchlets formed of other calicles.

Fig. 2. Ventral view of the same specimens; one of which is entirely bare of calicles on this side, and on the other only a few are present.

Fig. 3. Transverse section of a branch, showing the polyp cavities, the central cavity, and the canals running in various directions. The concentric arrangement of the latter is well shown. Into the central cavity project the six septa. In two of the innermost ring of polyps, the axial and abaxial septa have fused into the septal columella-plate.

Fig. 4. Diagram of a transverse section of a polyp and of part of the branch. The external body wall is shown to be supported on the costæ, as its mesoderm and endoderm are continuous with those lying on the outer face of the corallum. The polyp cavity shows at this point twelve mesenteries supporting the stomatodæum. (In nature the mesoderm lies closely apposed to the surface of the corallum, and there is no space between them, such as is introduced into the diagram for clearness.)

Fig. 5. Transverse section of a portion of the branch, to show the lines of growth running between the canals radially and terminating each in a costa.

Fig. 6. Diagram of a longitudinal section of a polyp along the dotted line in Fig. 8. The tentacles are omitted, as they were not recognisable in my specimens; the canal system in the corallum is also omitted. On the left the section passes between the axial septum and mesentery No. 7, and above the polyp down an external longitudinal canal; on the right, through the abaxial septum and down a costa, of which the echinulations and the canals between them are shown. The numbers indicate the same mesenteries as in Fig. 8. On the mesentery 7 is figured the endodermal swelling, with the bent canal indicated by dotted lines. In the stomatodæum are shown the two openings of the canals of mesenteries 7, 9, 11; and below the stomatodæum the free edges of these three mesenteries alone appear, the others dying out before this plane is reached. The dotted line indicates the junction of theca and septa, and the discontinuous character of the septum ($Ab.$) is clearly shown.

Fig. 7. The external body wall viewed from the exterior; the lighter spots are the places where the echinulations of the costæ have pierced the body wall on account of its shrinkage. This drawing shows the arrangement of the external longitudinal canals, and their connections between the spikes of the costæ. (Camera lucida.)

Fig. 8. Diagram of the various forms and conditions of the mesenteries in a polyp of Type A. Those numbered 1, 3, 5, 8, 10, 12 are unmodified and normal. The others, 2, 4, 6, 7, 9, 11, are modified in all the polyps of this type; they are from camera lucida drawings of the same mesentery at different heights. The arrows and Roman numerals in Fig. 6 show the planes in which the successive sections are taken; 2 shows the endodermal swelling, and the upper opening of the canal; 6 shows the lower opening; 9 is below the stomatodæum, and bears a filament; and in 11 no trace of the modification remains, the mesentery being normal, and similar to those of Type B.

Fig. 9. Transverse section of a modified mesentery, passing through both arms of the canal.

Fig. 10. Transverse section of the corallum of a branch of *M. aspera*.

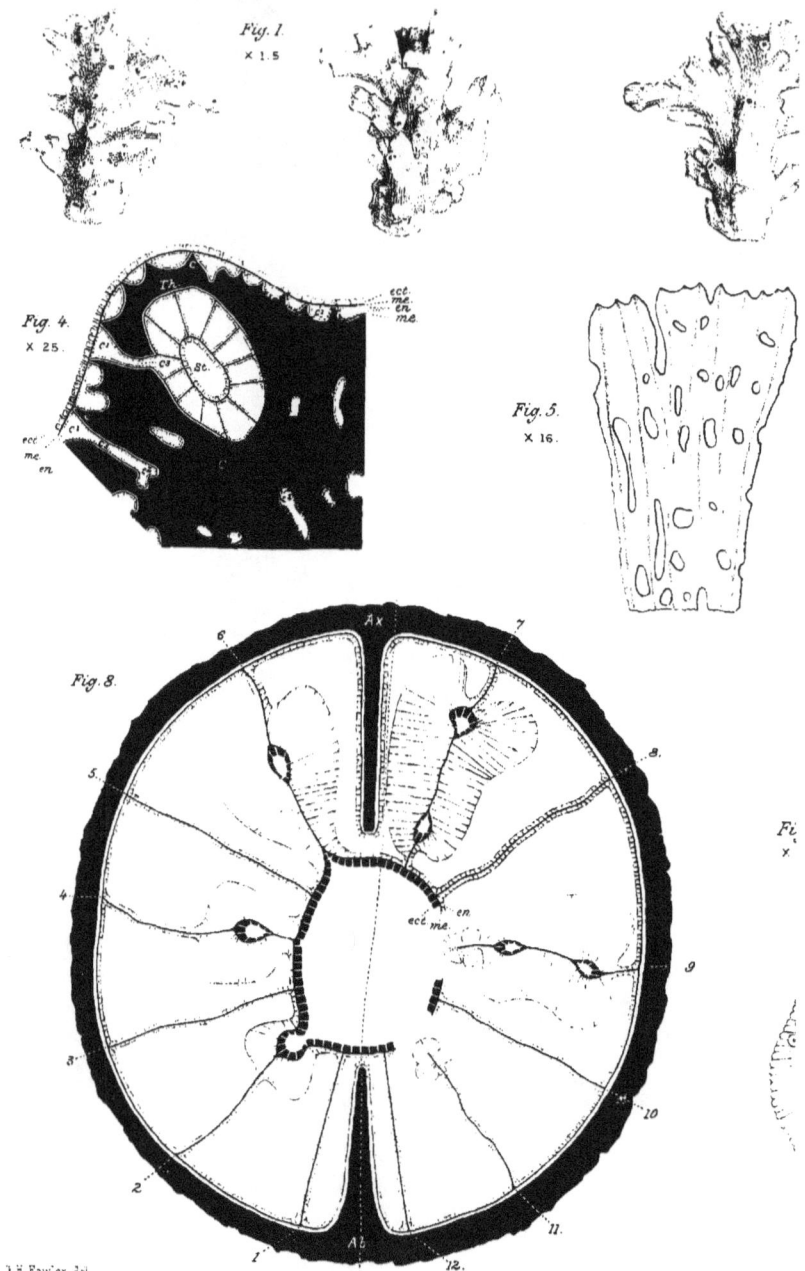

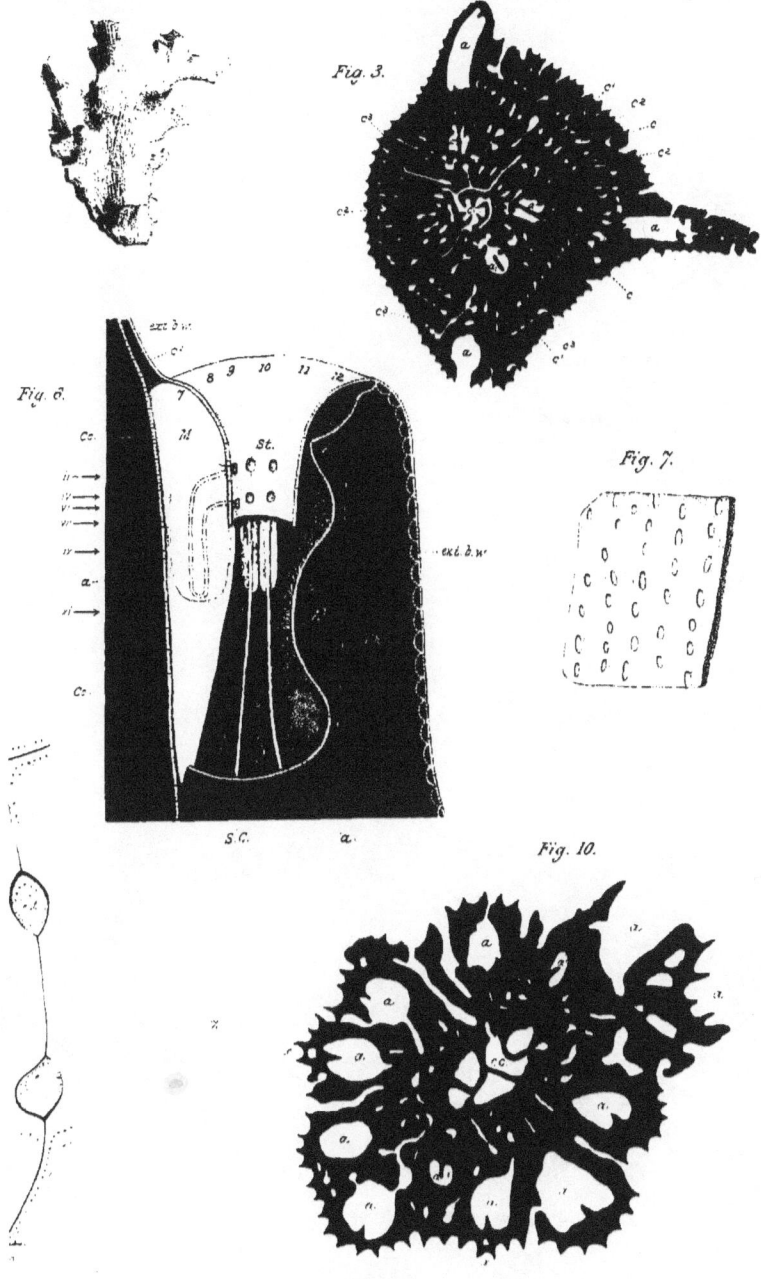

THE ANATOMY OF THE MADREPORARIA.
III.

By G. HERBERT FOWLER, B.A., *Keble Coll. (Oxon), Berkeley Fellow of the Owens College, Manchester.*

[WITH PLATES II. AND III.]

THE present memoir deals with the anatomy of *Turbinaria* (p. 17), a colonial Perforate coral; of *Lophohelia* (p. 22), an Imperforate form, colonial but with separate calyces; and of the two aberrant Imperforate genera, *Seriatopora* (p. 25) and *Pocillopora* (p. 28), in which the calyces are merged in cœnenchyme. To these descriptions is appended a note on the skeleton of Flabellum.

The most important facts now described for the first time are—1. The absence of directive mesenteries in *Lophohelia*, which thus differs from all Hexactiniæ hitherto described. 2. The retraction of the tentacles of Seriatopora by introversion, of which no other instance is known among the Madreporaria. 3. The presence of centres of calcification in the theca.

As in previous memoirs (2, 3), I have endeavoured to let the figures speak for themselves rather than to give detailed descriptions of structure.

TURBINARIA, sp. (Figs. 1–3).

For the opportunity of investigating this form, as in previous instances, I am indebted to the liberality of my teacher, Professor H. N. Moseley, who procured the material during the voyage of H.M.S. "Challenger."

I. CORALLUM.—The colony is crateriform or goblet shaped, the calyces of the polyps being placed on the inner face and on the brim of the goblet. The cœnenchyme is porous, in the manner characteristic of the Perforata, but the echinulations are not arranged in costæ with the regularity observable in some genera, except on the actual thecæ of the polyps. The latter project outwards from the cœnenchyme only abaxially, i.e., inwards towards the centre of the goblet, the axial half being almost level with the general surface. Specimens of the coralla of this genus are not uncommon in museums; a detailed description and figures are therefore unnecessary, and may be found in the works of the authors appended below (p. 22).

Owing to the small amount of the material at command, none could be spared for the determination of the species. It appeared, however, to belong to the type of *T. mesenterina.*

The septa, in fully-grown polyps of this particular species, vary much in number, but are generally from seventeen to twenty-two; they are entocœlic only. It is worthy of remark that the number of septa appears to bear no relation to any multiple of six, nor can any division into orders be effected, since all are approximately of the same length. A loose and incomplete columella, occurring deep down in the calyces, appears to be referable to fusion of the septa.

Part of a transverse section through the corallum (made according to the balsam and ether method introduced by von Koch) is represented in Fig. 1, showing sections through at least five polyp cavities. Of these one, a, is cut obliquely, owing to the sharp angle at which the polyp cavities are inclined to the general axis of the colony; of the others, which are cut through at varying distances from their orifices, that lettered b is a nearly transverse section, of a typical character, exhibiting eighteen septa; while the three others, c, show the reduction of the septa in the deeper parts of the cavities. The upper part of the figure represents the abaxial, the lower the axial, surface of the crateriform colony. The echinulations and canal system are also well shown in this section.

II. ANATOMY.—The whole colony, both inside and outside of the goblet, is clothed with an external body wall of ectoderm, mesoglœa,* and endoderm, exactly as has been described in Stylophora (7),

---

* The substitution of this word for the misleading "mesoderm" we owe to Bourne (1).

Madrepora (3), &c.; its continuity is only broken by the mouth orifices of the polyps.

What the exact relation of this body wall to the tissues actually apposed to the theca may be, *i.e.*, whether it agrees with the undoubted relations described in Astroides embryo (8), in Dendrophyllia (5), and Rhodopsammia (2), or with the apparently equally accurate relations recorded for Stylophora (7) and Madrepora (3), is exceedingly difficult to determine. In my specimens, both of Madrepora and Turbinaria, the contraction produced by preservation in alcohol has forced the body wall so tightly upon the echinulations that they project in many cases through it. Again, in Heteropsammia multilobata, a form as closely allied to Rhodopsammia as one with cœnenchyme can be to one devoid of it, the relations appear to be identical with those in Madrepora and Stylophora, and such as I have here (Fig. 2) drawn for Turbinaria. If we are justified in crediting the appearance of the tissues in Stylophora, Madrepora, and Turbinaria, which implies that the body wall is supported upon the echinulations (Fig. 2), it is not perhaps too much to infer that these relations are of *secondary significance, and have arisen contemporaneously with the development of cœnenchyme, for the support of the external body wall*, owing to the inadequacy of the peripheral sections of the mesenteries to effect this support elsewhere than immediately round the theca (where this is exsert from the cœnenchyme). In other words, the mesenteries are necessarily confined to the polyp cavity, and their peripheral sections to the small part of it which is cut off (?) from the rest by the upward growth of the theca; while therefore they are amply sufficient for the support of the body wall in a form with separate free calicles (*e.g.*, Rhodopsammia), they could not extend over the cœnenchyme of a form with fused or sunken calicles (*e.g.*, Heteropsammia multilobata), which is, by its very nature, outside of, and in a manner independent of, the polyp cavities.

Of the more primitive (?) condition, Astroides embryo (8), Rhodopsammia (2), Dendrophyllia (5), and Fungia (1) stand as admitted examples among the Perforata; Cladocora (4), and Caryophyllia (6) among the Imperforata, all being forms with free calyces; while of the secondary condition, Madrepora (4), Heteropsammia multilobata (which I hope to describe in a future memoir), Turbinaria among the Perforata, and Stylophora (7), Seriatopora and Pocillopora (described below) among the Imperforata, are the recorded instances, all possessing well-developed cœnenchyme.

In a minor point only do my observations differ from those of von Koch (7), viz., that he figures no mesoglœa between the echinulations and the ectoderm; in other words, according to his figure the persistent ectoderm of the body wall is at those points continuous with the calycoblast layer (Fig. 5). The reason which leads me to believe in the existence of a mesoglœa lamina between calycoblasts and external ectoderm, is that at the points where through shrinkage the echinulations have pierced the external body wall, they have carried with them this mesoglœa, which in sections of decalcified specimens preserves accurately their outline, projecting far beyond the shrunken ectoderm.

Between this external body wall and the corallum lies the system of approximately longitudinal canals with transverse commissures; in other words, the space between the body wall and the theca is broken up into canals by the points of contact. These canals communicate, as is usual in Perforata, with the canals which permeate the corallum and run also into the polyp cavities.

The polyps are built on the normal Actinian type. As the calyces are placed only on the inner side and on the lip of the crateriform colony, an easy identification of bilaterality is thus afforded, the dividing plane being a radius directed to the centre of the goblet. Approximately at the ends of this dividing plane are placed the axial and abaxial pairs of "directive" mesenteries, distinguished by the arrangement of retractor muscles on their ectocœlic faces. The polyps are not, however, rigidly bisymmetrical, inasmuch as the pairs of mesenteries lying right and left of the dividing plane are not equal in number.

The total number of pairs of mesenteries is not constant, but does not appear to depend upon the size (age) of the particular polyp. It varies generally from 17 to 22. The asymmetry of the polyps can best be seen in a tabular form:—

	A	B	C
Number of pairs of mesenteries	17	20	22
Number on right side of "directives"	7	10	9
Number on left side of "directives"	8	8	11

The three polyps here quoted were within a few millimetres of each other, and all were nearly of the same size.

The number of pairs of mesenteries is of course the same as that of the septa, the latter being entocœlic only, though a misleading appearance of ectocœlic septa is produced by the fact that some pairs of mesenteries die out after a very short course, while their septa are still recognisable at a much greater depth in the polyp cavity. The mesenteries with a longer course are in all respects perfectly normal, and in my specimens bore huge ova, the structure and relations of which call for no special comment (Fig. 3). The length or shortness of the mesenteries appears dependent on no particular system, such as has been observed in some other forms (3).

The tentacles are probably entocœlic only, but are so retracted as to render the point somewhat obscure. In this condition they are covered by a ring-fold formed of the indrawn margins of the disc, a method of protection common among the Actiniaria.

The histology, though much spoilt by prolonged decalcification, agrees with that of the typical forms already described. The muscular pleats of the mesoglœa of the mesenteries are only very slightly and irregularly developed, but entirely normal. Nematocysts are closely packed together in the tentacles; they are not, however, arranged in knobs or "batteries."

Zooxanthellæ are present abundantly in the canals exterior to the theca, in the tentacle cavities, and immediately under the mouth disc; elsewhere they are comparatively rare.

III. SUMMARY.—The following are the most important points elucidated:—

1. The polyps are of the normal *Actinian type*, and are *bilateral, but not rigidly bisymmetrical*.

2. The septa and tentacles (?) are entocœlic only.

3. The number of septa present is *inconstant*, and bears no relation to any multiple of six.

4. The general body wall of the colony is *supported upon the echinulations of the cœnenchyme;* a condition which *may* be of secondary significance, acquired for the purpose of such support, *contemporaneously with and in consequence of the development of cœnenchyme*.

IV. MEMOIRS REFERRING TO THE GENUS :—

MILNE-EDWARDS and HAIME, "Hist. Nat. des Coralliaires," iii., 164, pl. E i, Figs. 1*a*, 1*b*.

KLUNZINGER, "Korallthiere des Rothen Meeres," ii., 50.

LOPHOHELIA PROLIFERA (Figs. 4–8).

The material for a study of this form was entrusted to me by Professor E. Ray Lankester, who had dredged it off Lervik, Stordoe, Norway, and whom I am glad to be able thus to thank for his generosity. Owing to the great density of its corallum, and the consequent damage to the tissues produced by prolonged decalcification in a strongly acid medium, the work has been long delayed. Part of the material had been placed directly in absolute alcohol; part was passed from corrosive sublimate through successive strengths of spirit to 90 per cent. alcohol. Both sets were in excellent preservation, but the latter method appeared to be preferable, as resulting in less shrinkage of the tissues.

I. CORALLUM.—Of all corals this is probably the most generally familiar, and requires here no systematic description. The theca, which terminates the branches of the corallum, is solid, as in all such Imperforata. The septa, which are exsert above the lip of the theca, are both ectocœlic and entocœlic, but are only irregularly arranged in orders. In a polyp with forty-eight septa, for instance, of which twenty-four are ectocœlic, the remaining twenty-four entocœlic septa are probably divisible into six primaries, six secondaries, and twelve tertiaries; but, as they all are approximately of the same length, this division is founded more on analogy than on distinctive differences. The total number of septa, which probably varies with the age of the individual polyp, is not necessarily a multiple of six or twelve.

Transverse sections of the corallum show, as has been recorded for other forms, *e.g.*, Cladocora (**4**), Caryophyllia (**6**), that a dark line, indicating its earliest formed part, runs down the centre of each septum, and may be termed a "centre of calcification." In addition to these lines, however, sections, so made that they shall just cut the extreme lip of the actual theca, exhibit other "centres of calcification" between the enlarged ends of the septa, *i.e.*, they lie *in the theca itself* (Fig. 4). In sections at a lower plane the centres of calcification in the theca and in the ectocœlic septa are found to have run into a continuous dark line (Fig. 5), which at a yet lower level is joined by those of the entocœlic septa. There are thus three separate centres of active coral secretion at three different levels.

From Fig. 5 it is also obvious that by far the greatest thickness of the coral is laid down peripherally, *i.e.*, by the calycoblasts of the extrathecal part of the polyp. About six-sevenths of the thickness of the theca is due to these calycoblasts, while the remaining seventh is formed by those internal to the theca.

II. ANATOMY.—In spite of the great length of the branch on which it is borne, the polyp is often comparatively short, measuring from 5 mm. to 20 mm.

As will probably prove to be the case in all the Imperforata with free calyces (*cf.* Cladocora (**4**), Caryophyllia (**6**), &c.), the polyp is so continued over the lip and outer side of the calyx as to form a covering for its exterior surface to a varying distance (the "Randplatte" of von Heider). In Lophohelia this continuation may extend for about 15 mm., or even more; often it measures much less, and in it the relations of the various body layers are such as have been already described in the forms referred to above; the part of the cœlenteron enclosed in this "Rand-platte" is divided up into ectocœlic and entocœlic spaces nearly corresponding to those inside the calyx by the peripheral lamellæ, which were at a former time continuous with the more central mesenteries, but which have been mainly cut off from them by the gradual growth of the theca upwards, though the continuity is maintained above the lip (Fig. 6). This explanation, due originally to Dr. von Koch, must undoubtedly apply to this and many other *adult* forms.

The general anatomical relations of the polyp, and its agreement with forms already described, are shown in the diagrammatic segment of a transverse section (Fig. 6). The "Rand-platte," the mouth disc, tentacles, and stomatodæum are all in accordance with the normal type. The cœlenteron of the living polyp is, as usual, lined by endoderm and mesoglœa, apposed directly (except for scattered calycoblasts) to the corallum. At the point, however, where the living polyp ceases, its cœlenteron is separated off from the cavity in the coral which it previously occupied by a plug of decaying (?) tissue, in which no cell-elements or organic structure are recognisable, except occasionally the remains of the mesogloea lamina of a mesentery. Into this the living tissues pass gradually.

The tentacles, which are both ectocœlic and entocœlic, *i.e.*, one over every septum, are knobbed, each knob being such a battery of nematocysts as has been described in Flabellum (2), Stephanotrochus (11), &c.

The mesenteries, which, like the septa, vary in number in different polyps, all bear retractor muscles on their entocœlic faces, *i.e.*, there are *no pairs of "directive" mesenteries* at the opposite ends of the long axis of the oval stomatodœum, thus differing from those of all other Hexactiniæ or Madreporaria yet described. The significance of this fact cannot, of course, yet be understood, as nothing correspondingly abnormal occurs in any other part of the polyp with which it might be correlated. As, however, the mechanical or other function of the directive mesenteries is itself not yet explained, the meaning of the variation from the common type is naturally not appreciable. The number of the pairs of mesenteries, like that of the septa, is not necessarily a multiple of six.

The only point in the histology that appears worthy of note is the great length of the calycoblasts, as compared with that of the other cell-elements. A group of them from the edge of a growing septum is represented in Fig. 7. Still more marked is the great length of these cells in Fig. 8, which represents a transverse section through the tissues at that point where the upward growth of the theca divides the mesenteries into a central portion within the calyx, and a peripheral portion outside of it. Here they measure as much as ·054 mm. The large plate of mesogloea in the centre of the figure is merely that which immediately overlies the lip of the calyx, and is cut in a direction parallel to its flattened surfaces, while the section passes nearly at a right angle to the other tissues. The point here figured is such a "centre of calcification" in the theca as has been already referred to (*vide* p. 22).

III. SUMMARY.—The most important facts thus obtained are :—

1. The polyps agree with the normal *Actinian type*, except *for the absence of "directive mesenteries."* They possess a well developed "Rand-platte."*

---

* It is, perhaps, unnecessary to coin an equivalent for this till its morphological value is better understood.

2. The septa and tentacles are both *ectocœlic and entocœlic*, the number of septa not being necessarily a multiple of six.

3. Three series of centres of calcification are recognisable in the skeleton, of which one *lies in the theca itself*, and the other two at the summits of the ectocœlic and entocœlic septa respectively.

IV. MEMOIRS REFERRING TO THE GENUS :—

MILNE-EDWARDS and HAIME, "Hist. Nat. des Corall.," iii., 116.
STUDER, Steinkorallen auf der Reise S. M. "Gazelle" gesammelt,
      Monatsb. Akad. Berlin, 1877, p. 631, pl. i., Fig. 8.
MOSELEY, "Challenger" Rep. Zool., ii., 178, pls. viii., ix.

SERIATOPORA SUBULATA (Figs. 9-13).

For the material for the study of this coral and of Pocillopora I again owe my thanks to Professor H. N. Moseley, who has already investigated the general anatomy of both forms (10). As, however, no structural details have yet been figured, and these somewhat aberrant forms are of great interest, no apology is necessary for a second account of them. The specimens of Seriatopora were obtained by Mr. Gulliver from Zanzibar.

I. CORALLUM.—The characteristic feature of the skeleton which caused both Seriatopora and Pocillopora to be ranked in the now abandoned group of Tabulata is the presence of tabulæ, *i.e.*, successive floors of coral, by which the living polyp shuts off its cœlenteron from the cavity it previously occupied, a condition the opposite to that described above in Lophohelia prolifera. The calyces are therefore nearly confined to the outermost part of the colony, and are not continued deeply into it, as was the case in Turbinaria. These shallow calyces project but slightly above the cœnenchyme, and at a very short distance below the orifice are divided into two halves by the fusion of the two larger septa. These two septa, the axial and the abaxial, are the only two that are developed to any extent, though traces of the other ten may be recognised in many cases [*cf.* the condition of Madrepora Durvillei (3)]. When all are present there are six entocœlic and six ectocœlic. It is perhaps more accurate to speak of the calyx as divided into two halves by the fusion of these septa than to regard the two chambers thus formed as special pits for the reception of the two longer mesenteries (**10**), since they are simply downward continuations of the conical cœlenteron, and mesenteries other than the two longer ones are sometimes attached to their sides. Other details of the skeletal structure do not especially bear on the anatomy of the polyps.

II. ANATOMY.—As was shown by Professor Moseley, Seriatopora is undoubtedly a Madreporarian, and is even more in accordance with the normal types than could be inferred without the aid of sections.

The whole of the colony is clothed in the customary body wall of ectoderm, mesoglœa, and endoderm (Fig. 9), which is supported on the echinulations of the cœnenchyme (*vide supra*, p. 19). The space between body wall and theca is broken up by these spines into a superficial series of canals (Figs. 9, 10, 13), which ramify over the cœnenchyme and place the polyp cavities in communication with each other, but do not, of course, extend into the corallum in the manner characteristic of the Perforata. The body wall is continuous with the mouth disc, and from the centre of the latter rises a slight hypostome, through which opens the stomatodæum. This latter is crucial in transverse section, the longer arms of the cross being in the dividing plane of bilaterality indicated by the axial and abaxial septa (Fig. 10).

The tentacles, which are twelve in number, being both ectocœlic and entocœlic, are simple evaginations of the cœlenteron, tipped with a terminal swelling, which is a single " battery " of nematocysts (Fig. 11). There is, I believe, no instance yet recorded of the occurrence among Madreporaria of the method of tentacular retraction which distinguishes Seriatopora, namely, that of *introversion* (Figs. 12, 13), the tentacles being invaginated in such wise that the battery is still pointed upwards. In Fig. 13 the ectocœlic tentacles are expanded, while the entocœlic are introverted, a condition not uncommon in my specimens. Probably owing to the minuteness of the polyp, no special muscular apparatus for effecting this retraction could be detected.

The mesenteries, which are twelve in number, are arranged in pairs* on the normal type. In the diagram (Fig. 9) they are numbered in the same manner as those of Madrepora (3); the two mesenteries marked 3 and 10 respectively are comparatively long, extending to the bottom of the polyp cavity, and possess the thickened edge known as a mesenterial filament; of the rest, those numbered 1, 5, 8, 12, though generally devoid of a "filamentar" thickening, are recognisable in transverse sections for some distance below the stomatodæum; while the others, 2, 4, 6, 7, 9, 11, are rudimentary, and are visible only in the highest sections. It is worthy of remark that the six rudimentary mesenteries last mentioned are those which in the one type of polyp of *Madrepora Durvillei* are pierced by a special ectodermal canal, and which in the other type of polyp of the same species, and in all the polyps of *M. aspera*, are distinguished from the remaining six by a greater length and the possession of a filamentar thickening; in other words, *of the total twelve mesenteries the six which in the one form are the best developed are in the other quite rudimentary.*

The histology agrees with that of the normal types. I have found no trace of generative organs in my specimens.

III. SUMMARY.—The interesting points in Seriatopora are:—

1. The polyps are *Actinian* in structure.

2. The septa, when all are present, and the tentacles, are both *ectocœlic* and *entocœlic*.

3. *The tentacles are retracted by introversion.*

4. The body wall is supported upon the *echinulations of the cœnenchyme*.

5. Of the twelve mesenteries, six (and more especially two of these) are of some length, and six are rudimentary; but those which here are well developed are, in the Madreporæ mentioned above, rudimentary, and *vice versâ*.

---

* Professor Moseley (10) states that in Seriatopora and Pocillopora the mesenteries "are not disposed in pairs with regard to the septa," and the remark appears in a misleading form in Professor Martin Duncan's "Revision of the Madreporaria" ("Journ. Linn. Soc. Zool.," Vol. xviii.), to the effect that "the genera differ from other Madreporaria in not having their mesenteries arranged in pairs." The original statement was correct, because the possibility of ectocœlic septa in a coral had not been demonstrated.

IV. MEMOIRS REFERRING TO THE GENUS :—

MILNE-EDWARDS and HAIME, "Hist. Nat. Corall.," iii., p. 311, pl. F 4, fig. 3.

KLUNZINGER, "Korallthiere des Rothen Meeres," ii., 69, pls. vii., viii.

AGASSIZ, "Nat. Hist. United States," iv., p. 296, pl. xv., fig. 15.

POCILLOPORA BREVICORNIS (Figs. 14, 15).

The anatomy of this species agrees so closely with that of *Seriatopora subulata* that only points of difference between the two need be quoted.

The corallum is, of course, different in its mode of growth, as upon this the distinction between the two genera is based, but this difference does not affect the anatomical relations of the polyps. The method of support of the external body wall is identical with that in Seriatopora; the tentacles agree in the two forms, though, as they are fairly well expanded in my specimens, it does not appear whether they are capable of introversion or not; the stomatodæum is less distinctly conical than in the cognate genus.

As regards the mesenteries, the only points of difference noticeable are, that in Pocillopora those denoted in the diagram (Fig. 9) by the numbers 3, 10, are not proportionately so much longer than those marked 1, 5, 8, 12, and that a mesenterial filament may sometimes be detected on the four last mentioned; in other words, *the tendency observed in both Seriatopora and Madr. Durvillei towards the exclusive assumption of function on the part of six mesenteries and towards a correlated retrogression (?) on the part of the other six, has not attained to such a pitch in Pocillopora (and Madr. aspera) as in the other two forms.*

The statement of Professor Moseley (10), that "the mesenterial filaments are not enclosed in prolongations of the chamber walls," is not justified by the examination of sections; the two longer and more developed mesenteries, with their filaments, 3, 10, lie, as in Seriatopora, for their whole length in the coelenteron.

Apparently, any of the mesenteries may bear generative organs, and it is worthy of remark that the polyps are *monœcious*. The ovaries and testes, though surrounded by a thin capsule of mesoglœa and endoderm, as in typical forms, do not lie, as is generally the case, in the plane of the mesenteries (*cf.* Fig. 3), but project from their sides in a manner more characteristic of certain Alcyonaria, so that in transverse sections of the colony they frequently appear to lie free in the cœlenteron. Two stages in the development of the spermatozoa are figured as well as the preservation of the material would allow (Figs. 14, 15).

In both this and Seriatopora there was left, after decalcification, a residue in the position occupied by the corallum, which, though staining faintly both with hæmatoxylin and with borax carmine, showed no distinctly organic structure.

In transverse sections of the polyps it is just possible to detect, at the point of the insertion of the mesenteries in the corallum, structures similar to those described by Sclater (11) as calycoblasts. Their excessive minuteness in Pocillopora rendered an accurate investigation impossible, but they certainly appeared to me to be rather connected with the attachment of the mesentery to the corallum than with the secretion of coral.

MILNE-EDWARDS and HAIME, "Hist. Nat. des Corall.," iii., p. 301, pl. F 4, figs. 1, 2.

AGASSIZ, "Nat. Hist. United States," iv., 295, pl. xv., 14.

KLUNZINGER, "Korallthiere Roth. Meer.," ii., 66, pls. vii., viii.

NOTE ON THE SKELETON OF FLABELLUM.

In his latest addition to the literature of the subject (9), the main part of which I do not propose at present to discuss, Dr. von Koch treats, amongst others, of the skeleton of Flabellum.

1. He states that the dark line of growth visible in transverse sections of the calyx, which indicates the earliest formed part of the coral at that level, is in Flabellum placed peripherally (Fig. 16), and consequently that the skeleton is laid down from without inwards.

2. Elsewhere in the same paper he infers, from his researches on the development of *Astroides calycularis* (8) that the epitheca of *all* corals, originally deposited outside the lateral body wall of the embryo, also increases in thickness on the inner side only.

3. Finally, we find that "diese Koralle bildet einen ganz eigenen Typus, wegen des gänzlichen Fehlens der Innen-platte. Die Aussen-platte* ist gut entwickelt . . . Die Homologisirung der Aussen-

---

* *i.e.*, Epitheca.

platte mit der Innen-platte (Theca) der vorhin beschriebenen Korallen wird aus der Struktur derselben als irrig erkannt."

The implied argument may thus be expressed in the syllogism :—

1. The skeleton of Flabellum grows in thickness from without inwards.

2. An epitheca grows in thickness from without inwards.

3. Therefore the skeleton of Flabellum is an epitheca—an example of what is characterised by logicians as the fallacy of the undistributed middle term ("Medium non Distributum").

Dr. von Koch is no doubt correct in asserting that the calyx of Flabellum is laid down from without inwards; but till clearer evidence be adduced to the contrary, it is far simpler to regard it as a theca entirely homologous with the theca of typical Madreporaria (or at least with a part thereof), than to conceive that the epitheca, which we know elsewhere only as an inconstant and inconsiderable structure, should have replaced the solid theca, *merely to achieve the same physiological end.*

Nor is there anything in the structure of the corallum really inconsistent with the idea that it is a theca. The embryonic *Flabellum patagonicum* attaches itself to an Arenaceous Foraminifer, or some similar body (*vide* Moseley, "Rep. Chall. Zool.," ii., Madrep., pl. xv., figs. 1, 2); but the adult is entirely free, and therefore more or less at the mercy of natural accidents such as currents. Correspondingly with this condition, but unlike that of the attached forms (Lophohelia, Caryophyllia, &c.) it develops no "Rand-platte," (*vide* p. 23), but the polyp can be wholly retracted within the calyx (*cf.* (**2**) Fig. 2). The absence of the "Rand-platte" implies almost necessarily the absence of extracalicular calycoblasts; the calyx must therefore be deposited by those internal to the corallum. As a consequence of these facts, the calyx of Flabellum, if it be not an epitheca, would be homologous with that part of the theca of Lophohelia, &c., which lies internal to the dark line of growth mentioned above (p. 22); and a comparison of Fig. 16 with Figs. 4, 5, will show that there is no discordance between the two structures.

In both forms, as is generally the case, the regions due to separate centres of coral secretion are bounded by sutures; and of these regions those marked T. (theca), and S. (septal), in all three figures certainly appear to be respectively homologous. Even the way in which the dark line in Lophohelia curves inwards to the ectocœlic septum (owing to the fact that at the lip the latter does not project so far peripherally as an entocœlic septum) agrees with the involution of the septa in Flabellum. The fact that in Fig. 5 the centre of calcification of the entocœlic septum projects outwards through the line of growth, is of course attributable to the pseudo-costæ occurring at the lip of the calicle of this species of Lophohelia which are produced by extra-calicular calycoblasts, and are not therefore represented in Flabellum.

In Fig. 17 is drawn a transverse section through a part of the pedicle, that is to say, a section through the corallum of an embryo Flabellum measuring about 2·25 mm. in diameter, and possessing six primary and six secondary septa. The relations indicated by the sutures are the same as in the former section. The successive laminæ showing the conversion of the embryonic calyx into a nearly solid pedicle are well marked.

In conclusion, I have to express my thanks to Professor Milnes Marshall for his assistance; and to the anonymous donor of the Berkeley Fellowships in the Owens College, whose generosity has enabled me to carry on my studies.

List of Memoirs Quoted.

1. Bourne, G. C.—"The Anatomy of Fungia," "Quart. Journ. Micr. Sci.," xxxvii.
2. Fowler, G. H.—"The Anatomy of the Madreporaria," I. "Quart. Journ. Micr. Sci.," xxv.
3. Fowler, G. H.—"The Anatomy of the Madreporaria," II. "Quart. Journ. Micr. Sci.," xxvii.
4. Von Heider, A.—"Die Gattung Cladocora," "Sitz. k. Akad. Wiss. Wien," lxxxiv.
5. Von Heider, A.—"Korallenstudien," "Arb. Zool. Inst. Graz.," i.
6. Von Koch, G.—"Bemerkungen über das Skelet der Korallen," "Morph. Jahrb.," v.
7. Von Koch, G.—"Mittheilungen über Cölenteraten," "Jen. Zeitschr.," xi.
8. Von Koch, G.—"Entwicklung des Kalkskelettes von Astroides calycularis," "Mittheil. Zool. Sta. Neap.," iii.

9. Von Koch, G.—" Ueber das Verhältnis von Skelet und Weichtheilen bei den Madreporen," " Morph. Jahrb.," xii.
10. Moseley, H. N.—"Notes on Seriatopora, Pocillopora, &c.," " Quart. Journ. Micr. Sci.," xxii.
11. Sclater, W. L.—"On a New Madreporarian Coral (*Stephanotrochus Moseleyanus*)," " P. Zool. Soc.," 1886.

EXPLANATION OF PLATES II. AND III.

Illustrating Mr. G. Herbert Fowler's Paper on " The Anatomy of the Madreporaria," III.

Fig. 1. Transverse section through a small part of the crateriform colony of *Turbinaria sp.* (*vide* p. 18). *Ab.* Abaxial, inner, or ventral surface of the colony. *Ax.* Axial, outer, or dorsal surface of the colony. *a.* Oblique section through a polyp cavity. *b.* Transverse section of a polyp cavity near its orifice. *c—c.* Similar sections further from the orifices. (Camera lucida.)

Fig. 2. Diagrammatic transverse section of a polyp of *Turbinaria sp.* (p. 19). In this and similar diagrams the ectoderm is represented by " blocked " black and white, the mesoglœa by a dark line, and the endoderm by a light line, the calcareous skeleton being dotted. The thicker and contorted ectoderm at the upper part of the stomatodæum represents the taller cells, interspersed with plentiful nematocysts, occurring in the neighbourhood of the tentacles, *i.e.*, mouth disc rather than stomatodæum. Twenty-two pairs of mesenteries occur in this polyp, of which eleven are on the right and nine on the left of the " directives." A bit of the external body wall is drawn to show its relations to the echinulations. *Ab.D.* Abaxial directives. *Ax.D.* Axial directives. *St.* Stomatodæum. *Ect.* Ectoderm. *Me.* Mesoglœa. *En.* Endoderm. (Cam. luc.)

Fig. 3. Transverse section of a mesentery of *Turbinaria sp.*, bearing an ovum with nucleus, nucleolus, and nucleoleoli. The lengthening of the endoderm cells round the ovum is noticeable.

Fig. 4. Transverse section of the calyx of *Lophohelia prolifera* near the lip (*vide* p. 22). The darker parts represent "centres of calcification," or the earliest deposited portions, which become enlarged into the regions marked respectively $T.$ (thecal), or $S.$ (septal), according to their origin. The regions are bounded by "sutures." *Ect. s.* Ectocœlic septa. *Ent. s.* Entocœlic septum.

Fig. 5. Similar section of *Lophohelia prolifera* at some distance from the lip of the corallum. The "centres of calcification" of the ectocœlic septa and of the theca have run into one line, owing to the growth of the latter upwards to the former. Lettering as in Fig. 4.

Fig. 6. Diagrammatic transverse section through a segment of *Lophohelia prolifera*. The septa are seen to stand in both ectocœlic and entocœlic spaces. The peripheral sections of these spaces and the peripheral lamellæ of the mesenteries, cut off by the upgrowth of the theca, are also apparent. Lettering as in Fig. 2.

Fig. 7. Tissue from the growing edge of a septum of Lophohelia, obtained by a longitudinal section of the polyp. *cb.* Calycoblasts. *me.* Mesoglœa. *en.* Endoderm.

Fig. 8. Tissue surrounding a thecal centre of calcification, obtained by a transverse section of the polyp (*vide* p. 24), showing the separation of a mesentery into central and peripheral parts in process. *Cœl.* Intrathecal cœlenteron. *Cœl'.* Extrathecal cœlenteron. $M.$ The central, and $M'.$ the peripheral part of the mesentery. Others letters as before. (Cam. luc.)

Fig. 9. Transverse diagrammatic section of a polyp of *Seriatopora subulata*. The mesenteries are numbered 1-12, in the same manner as the Madreporæ before described (3). Letters as in Fig. 2. (This diagram is also good for Pocillopora.)

Fig. 10. View of a polyp of Seriatopora from above. The clearer spaces in the body wall of the colony represent the positions of the echinulations on which the body wall is supported, they having been dissolved away by acid. Through the body wall are seen the pair of longer mesenteries, 3 and 10. (Cam. luc.)

Fig. 11. Longitudinal section of a partly expanded tentacle of Seriatopora, showing the single battery of nematocysts, at the tip, interspersed with a few deeply staining gland cells.

Fig. 12. Diagram of a longitudinal section of an introverted tentacle of Seriatopora. *B.* The battery of nematocysts, pointed upwards.

Fig. 13. Diagram of an ideal longitudinal section through a polyp of Seriatopora, along the line *a...a* in Fig. 9. Of the six tentacles, three are expanded and three are introverted, one of the latter being cut longitudinally. Of the mesenteries, that on the right of the figure (10) is one of the two longest; that on the left (5) is much shorter; while 6 and 7 are rudimentary, and do not reach as far as the end of the stomatodæum. The cavity is divided into two halves by fusion of the axial and abaxial into one median septum.

Fig. 14. Early stage in the development of spermatozoa in Pocillopora.

Fig. 15. Later stage of the same. The testis is surrounded on all sides by endoderm, owing to the projection of the capsule outwards from the plane of the mesentery (*vide* p. 28).

Fig. 16. Transverse section through the calyx of Flabellum (*vide* p. 30). The numbers ii., iii., iv., indicate the orders to which the septa respectively belong. Other letters as Fig. 4.

Fig. 17. Transverse section through part of the pedicle of Flabellum, showing the conversion of the embryonic theca into a nearly solid pedicle. i., ii., Primary and secondary septa.

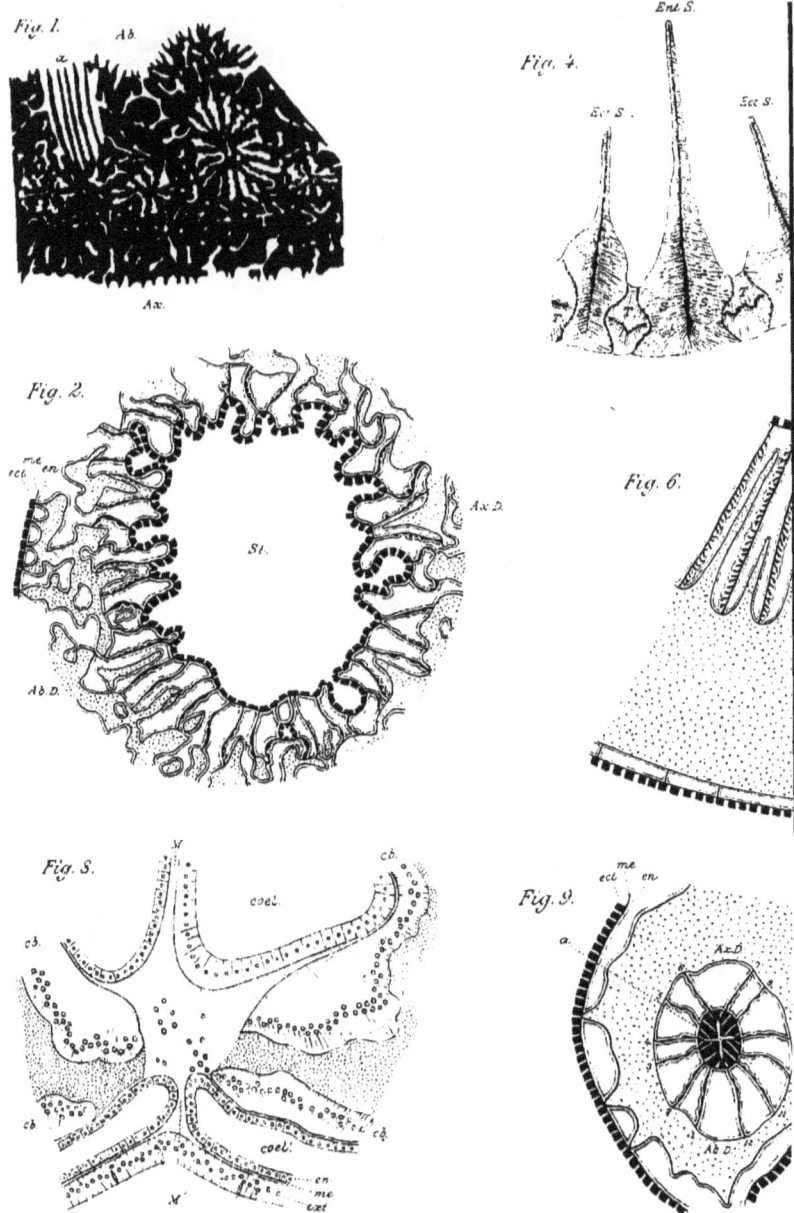

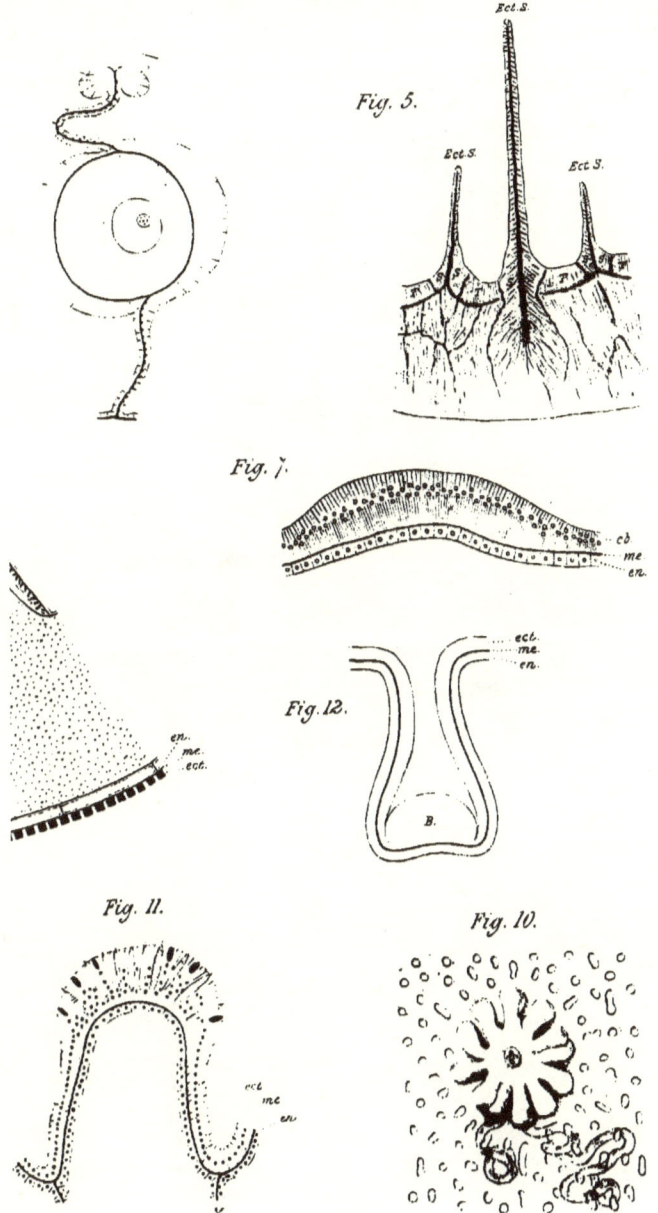

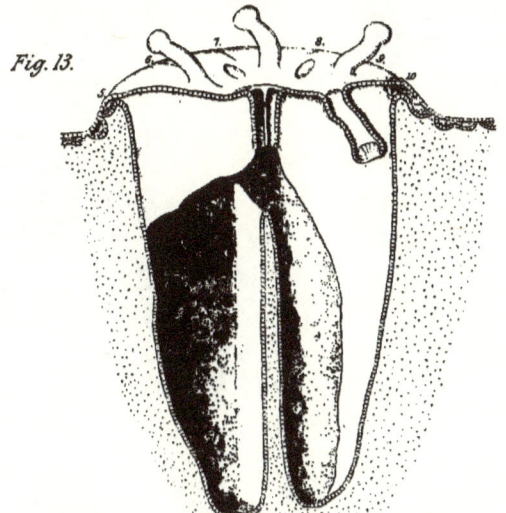

Fig. 13.

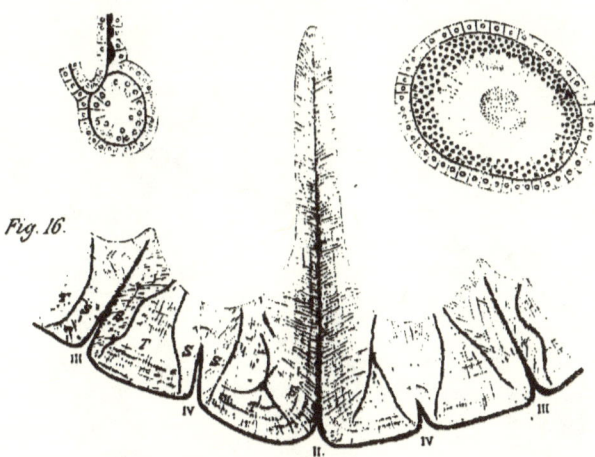

Fig. 14.   Fig. 15.

Fig. 16.

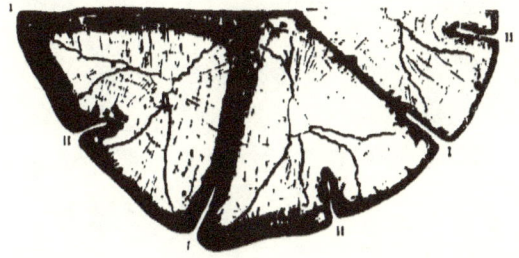

Fig. 17.

ON THE POSITION AND PERITONEAL RELATIONS OF THE MAMMALIAN OVARY.

By ARTHUR ROBINSON, M.B., C.M. Edin., *Demonstrator of Anatomy in the Owens College.*

[PLATE IV.]

THE mammalian uterus is either double, as in monotremes and marsupials; single, as in monkeys and man; or it consists of a body and two diverging cornua, as in the majority of mammals.

It is always attached to the abdominal wall by means of a peritoneal fold called the broad ligament. In the case of the bicornuate uterus, each cornu is attached to the dorsal wall of the abdomen by a broad ligament, which extends forward beyond the extremity of the uterine cornu as far as the diaphragm.

The broad ligament consists of two layers of peritoneum, between which the vessels and nerves of the uterus, the Fallopian tube, and the ovary are placed.

The dorsal edge of the fold is attached to the inner surface of the dorsal wall of the abdomen, and, along the line of attachment, the two layers become continuous with the general peritoneal lining of the abdominal cavity. The other edge of the fold is placed ventrally, and hangs free in the abdominal cavity. One surface of the ligament looks toward the mesial plane of the body, the other toward the lateral wall of the abdomen. It may, however, be modified, and be shortened until it is practically limited to the pelvis, as in the human female.

The ovary in many mammals is placed immediately posterior to the kidney, and is attached to the mesial face of the broad ligament. Nevertheless it often passes backward, and in the human female is normally situated at the brim of the pelvis.

In all cases, however, whether it is placed far forward in the abdominal cavity or has receded to the pelvic brim, it is attached to that face of the broad ligament which was originally internal.

As a general rule the ovary is more or less ovoid in shape, and is compressed from side to side, so that it has two surfaces, two borders, and two extremities. Its surfaces may be grooved, as in *Beluga catodon*,* or smooth, as in *Cœlogenys pacu*, guinea pig, antelope, &c., rough and tubercular, as in the sow;† or it may consist, as described and figured by Owen in the Wombat,‡ of a number of almost separate follicles.

By one edge it is attached to the mesial surface of the broad ligament; the other edge is free, and is directed toward the ventral wall of the abdomen, consequently one surface of the ovary looks outward, and is in relation with the inner face of the broad ligament, while the other is directed inward. The long axis of the ovary is parallel to the long axis of the body, one extremity being anterior, the other posterior.

A white streak, which runs along the margin of attachment to the broad ligament, marks each surface of the ovary; it has been called the "white line," and it indicates fairly accurately the region of transition of the germinal cells covering the free surfaces of the ovary into the endothelium lining the inner surface of the peritoneum. Two strands of fibrous tissue, both connected with the ovary, are placed in the mesial layer of the broad ligament: one extends from the posterior extremity of the ovary to the tip of the cornu uteri, the other extends forward from the anterior extremity of the ovary, and is either lost in the broad ligament on the outer side of the kidney, or it passes forward to the diaphragm. The posterior of the two strands is called by the human anatomist the ovarian ligament, a term which it is convenient to retain. The anterior strand has been called the diaphragmatic ligament. The two portions correspond

---

* M. Watson and A. H. Young, *Trans. Roy. Soc. Edin.*, vol. xxix., p. 431.
† Owen, *Anat. of Vert.*, vol. iii., p. 694.
‡ *Loc. cit.*, vol. iii., p. 681.

respectively to the posterior and anterior portions of the ovarian ligament, described by Owen in the Ornithorhynchus.*

Both strands give rise to reduplications of the mesial layer of the broad ligament; and between these reduplications, together with the ovary on the inner side, and the broad ligament on the outer side, the peritoneum is depressed into a shallow pouch, the mouth of which is widely open ventrally. This pouch may become developed into a distinct sac, and it will be convenient to speak of it afterwards as the "ovarian sac."

The outer wall of the sac is greater in vertical height than the inner, usually exceeding it to such an extent that the free edge of the ovary, which projects beyond the inner edge of the sac, is on a level with the free edge of the broad ligament.

The relations of the Fallopian tube and its abdominal orifice to the sac are of considerable importance, and it becomes necessary therefore to consider their position carefully.

The Fallopian tube is found to be placed between the layers of the broad ligament, usually a little dorsal to its free edge, though in many cases, and among others the human female, the cat, and the tiger, it occupies the free edge of the ligament. It is, however, always placed in the outer wall of the sac. Its orifice is situated either on the inner face of the outer wall of the sac, from which it projects slightly, or on the free edge of the broad ligament, from which it also projects. The abdominal termination of the oviduct is funnel-shaped; the funnel is compressed from above downwards, consequently the orifice is an elongated slit. The anterior extremity of this slit is attached to the anterior extremity of the ovary; the posterior extremity remains free. The orifice lies on the outer edge of the sac, and is parallel with the ovary. This simple form of sac is of common occurrence, and it has been described and figured by various anatomists. It is met with in *Hyæna crocuta*,† in the rabbit, cat, tiger, &c. It becomes modified and rendered more distinct in the following manner:—The Fallopian tube becomes elongated, its orifice being fixed, the tube is pushed forward between the layers of the broad ligament, and becomes folded on itself. This increase in length of the Fallopian tube is accompanied by a simultaneous

---

* *Loc. cit.*, vol. iii., p. 677.
† M. Watson, *Proc. Zool. Soc.*, May 1, 1877, p. 372.

growth of the broad ligament, the increase specially affecting that portion of the ligament which lies ventral to the attachment of the ovary. As a result a deep sac is formed, which extends forward beyond the ovary and hangs down below it. The orifice of the sac tends to be turned toward the mesial plane of the body, and the ovary occupies the upper edge of the orifice of the sac, while the Fallopian tube opens on the lower edge. An example of a pouch thus formed is found in *Cœlogenys paca;* but a still better marked example is presented by *Cynocephalus porcarius.*

The direction of the orifice of the sac depends upon the lengthening of the Fallopian tube, and upon the increase in extent of the broad ligament lying ventral to it. If the increase of the peritoneal fold is comparatively great the sac hangs down below the ovary, and the orifice of the sac looks upward and inward towards the vertebral column, as described by M. Watson in the Indian Elephant.* If at the same time the Fallopian tube is much elongated, it forms folds and convolutions in the outer wall of the sac.

The next step in the completion of this ovarian sac is brought about by the adhesion of the edges of the adjacent folds, as follows: The free edge of the broad ligament, between the uterus and the posterior extremity of the orifice of the oviduct, becomes adherent to the edge of the fold caused by the ovarian ligament.

In this manner the pouch is rendered more sacciform, and its opening into the peritoneal cavity is reduced to a mere slit, equal to or slightly exceeding in length the long diameter of the ovary. On the upper edge of the slit the ovary is placed; on the lower edge the orifice of the oviduct is found. A good example of a pouch thus formed may be found in the guinea pig, in which animal the Fallopian tube is folded and convoluted in the outer wall of the sac, and another in the porcupine. Ovarian peritoneal pouches, similar to those above mentioned, have been often figured and described. They have been found in the seal,† in *Hyæna crocuta,*‡ in the wombat,§ in the rabbit, the guinea pig, bitch, in insectivores,‖

---

* *Trans. Zool. Soc.*, vol. xi., part iv., p. 114.
† Turner, *Trans. Roy. Soc. Edin.*, vol. xxvii., p. 277.
‡ M. Watson, *Trans. Zool. Soc.*, vol. xi., p. 114.
§ Owen, *Anat. Vert.*, vol. iii., p. 680.
‖ Owen, *loc. cit.*, vol. iii., p. 688.

in *Bradypus tridactylus*,* in *Globiocephalus melas*,† and *Rhinoceros indicus*,‡ &c., &c.

The way in which the orifice of the sac becomes still further diminished in extent has not, so far as I am aware, been described, though the sacs have been mentioned explicitly by Owen, who describes a very complete sac in the case of the white bear. The steps of the process appear to be the following :—As the sac increases in relative size, from growth of its walls, the edge of the orifice of the sac, in which the oviduct opens, overlaps the ovary, and finally conceals it. When this growth of the sac, and consequent change of position of its orifice has taken place, the ovary is suspended from the broad ligament, and lies within a peritoneal capsule, the only opening of which is situated on the mesial side of the ovary, along its line of attachment to the inner face of the broad ligament, and now the orifice of the Fallopian tube lies along the inner margin of the opening of the sac.

The further closure of the pouch is brought about by the attachment of the edge of the orifice of the oviduct to the inner side of the ovary along the white line. The adherence takes place from before backward; a very peculiar appearance is the result. If the Fallopian tube is traced from the end of the cornu uteri, it is seen to extend along the outer side of the ovary to its anterior extremity; at this point it turns back, and, running along the inner side, terminates at a minute orifice situated between the posterior extremity of the ovary and the tip of the cornu uteri.

The minute orifice opening into the peritoneal cavity does not by any means form the whole of the abdominal termination of the oviduct, but only that portion of it which still remains unattached either to the ovary or to the opposite edge of the capsule. The greater part of the abdominal termination of the Fallopian tube is enclosed in the ovarian capsule, the formation of which has already been described.§

---

* Owen, *loc. cit.*, vol. iii., p. 690.

† Murie, *Trans. Zool. Soc.*, vol. viii.

‡ Owen, *loc. cit.*, vol. iii., p. 693.

§ Between this almost complete closure of the sac and the slit-like opening found in the porcupine and guinea pig, intermediate stages are presented by the bitch and the wolf.

Examples of this stage of completion of the sac are presented by the common badger and the racoon (*Procyon lotor*).

In both these animals the only opening from the Fallopian tube into the peritoneal cavity is extremely small, and is situated between the posterior extremity of the ovary and the tip of the cornu uteri on the inner face of the broad ligament. Only a very small portion of the orifice of the Fallopian tube presents itself here, the orifice being really enclosed in the ovarian sac, and it would be more correct to call the minute opening above described an opening of communication between the general abdominal cavity and the ovarian sac; from it, however, one or two fimbriæ of the infundibulum project.*

So far we have noted four types or classes of peritoneal capsules connected with the ovary; they are as follows:—

1. A class in which the pouch is shallow and widely open ventrally; the ovary is situated on the inner side of the pouch, and the orifice of the oviduct is on the outer side. The oviduct runs almost directly from the uterus to its abdominal termination, forming a very slight curve. Examples of this class are—the rabbit, tiger, antelope, cat, hyæna, koala, ornithorhynchus, sow, &c.

2. A class in which, on account of the elongation of the Fallopian tube and growth of the broad ligament, a more distinct pouch has been formed, hanging down below the ovary. The orifice of the pouch is wide, and is directed toward the mesial plane. The opening of the oviduct is situated on the lower border of the orifice

---

* Among a series of female generative organs, belonging to the Owens College, kindly placed at my disposal by Professor Young, I found the generative organs of the racoon previously described and figured by the late Dr. Watson in the *Proceedings of the Royal Society*, No. 213, p. 2. After describing the course of the Fallopian tube round the ovary, and the minute peritoneal opening situated between the posterior extremity of the ovary and the tip of the uterine cornu, Dr. Watson stated that the ovary was devoid of any peritoneal capsule, and that there was no infundibulum to the Fallopian tube. At first sight I found a state of affairs just like that figured and described in the paper quoted, but on closer examination of the parts, made with the light thrown upon them by the examination of other animals, it was evident that what at first sight appeared to be the surface of the ovary was in reality only the wall of the capsule, which in this animal, instead of being lax, as is often the case, is drawn tightly over the face of the ovary, and corresponds with its eminences and depressions. When the sac wall was opened along the front by an incision, the ovary was found in the position already described. The infundibulum of the Fallopian tube was enclosed in the sac lying close to the inner face of the ovary along its line of attachment, to the broad ligament a few fimbriæ only projected from the minute opening in the capsule by which the ovarian sac communicated with the peritoneal cavity.

of the sac, the ovary is on the upper border, and projects down into the sac. The Fallopian tube forms a distinct widely open curve in the wall of the sac. Examples—*Cœlogenys paca, Cynocephalus porcarius*, elephant, &c.

3. A class in which the growth of the Fallopian tube and broad ligament takes place as in the second class, but the edge of the broad ligament posterior to the ovary has become adherent to the free edge of the secondary fold caused by the ovarian ligament; consequently the orifice of the sac, which is directed inward and upward towards the spine, is a narrow slit, equalling, or but slightly exceeding, in length the long diameter of the ovary. Examples—guinea pig, bitch, porcupine.

4. A class the sac of which is formed as in the third class, but, further, the lower edge of the orifice has become adherent to the inner face of the ovary along the white line, except at the point posteriorly where a minute orifice is left, from which one or two fimbriæ project. The ovary and the greater portion of the infundibulum are enclosed in the sac, and are not visible until the sac wall has been opened. Examples—racoon and badger.

In both the examples of the latter class it must be noted that the ovarian ligament is very short, almost absent. The ovary lies close to the tip of the uterine cornu, the Fallopian tube is short, and the sac wall not flaccid, but closely applied to the surface of the ovary.

Numerous examples of these different types, or forms, of ovarian sac have been described and figured by previous observers. There is, however, a still further modification of the relations between the cavity of the ovarian sac and the general cavity of the peritoneum, and one of peculiar interest, since it results in the entire separation of these cavities from each other, and therefore in the formation of a distinct and complete ovarian sac. This condition I have met with in animals so common as the rat and mouse; still, so far as I am aware, it has not yet been described, and I have therefore the less hesitation in referring to it in detail.

While examining the reproductive organs of the white variety of the common house mouse (*Mus musculus*), and also those of the white variety of the common brown rat (*Mus decumanus*), I found that on superficial observation the ovaries appeared to lie outside the peritoneum; but when the layer of peritoneum covering the surface of the ovary was carefully lifted up a sac was at once demonstrated, in which the ovary lay, and on the wall of which the Fallopian tube formed many convolutions. No opening into the sac was visible.

In order that no small orifice might escape observation, a series of animals was taken, and through the uterus and Fallopian tubes injections, in some cases of coloured fluids, in others of air, were forced. These injections entered and distended the ovarian sacs, but in no case escaped from them into the peritoneal cavity. Thus it was definitely proved that the only communication of the sac was with the uterus through the Fallopian tube, and that there was no opening from the sac into the peritoneum.

As further proof, several ovaries were cut out, both from rats and mice, together with the broad ligament, the tip of the cornu uteri, and the Fallopian tube in each case. These specimens were stained with borax carmine embedded in paraffin, and a continuous series of sections was taken from them by means of the Cambridge rocking microtone.

When the sections were examined, it was found that the ovary lay in a sac (Plate IV., Fig. 5), the wall of which became continuous with the ovary and the broad ligament at the hilus of the ovary.

The sac wall consisted of fine fibrous tissue : both surfaces of the wall were covered with flattened, plate-like cells. The cells on the outer surface of the sac wall were continuous with the general epithelial lining of the peritoneal cavity, and were in every way similar to them. Those on the inner surface of the sac over its greater portion were also similar in character, but as they approached the hilus of the ovary their form began to change, becoming more and more cubical, until by gradual transition they became transformed into the germinal cells covering the free surface of the ovary.

The wall of the ovarian sac was complete, except at the point of entrance of the Fallopian tube. At this place the sac wall became continuous with the outer wall of the Fallopian tube, the dilated termination of which projected into the interior of the sac.

There was thus demonstrated a sac completely surrounding the ovary, shut off from the peritoneal cavity, and communicating at one place with the lumen of the Fallopian tube.

The same arrangement of parts was found present in rats three days old, the only difference from the adult being as follows :—The sac wall was thicker and more closely applied to the surface of the ovary; the cavity of the sac was less distinct, and the fimbriated orifice of the Fallopian tube lay close along the inner face of the ovary.

Thus the rat and the mouse present us with a fifth type or class of peritoneal ovarian sac, namely, a sac the cavity of which is completely shut off from the abdominal cavity, and whose only communication is with the uterus, through the Fallopian tube.

In most mammals the ovary is placed in the abdominal cavity, or in a pouch of peritoneum widely open to the abdominal cavity, and the ova, as they escape from the ruptured follicles, are swept into the open extremity of the Fallopian tube. In the rat and mouse, however, the ovary is placed in a sac, the cavity of which is undoubtedly, originally, a portion of the abdominal cavity; but in the adult animal, indeed, in the case of the rat and mouse, from the period of birth the cavity of the ovarian sac is, as before stated, completely shut off from any communication with the abdominal cavity. The ova pass from the ovary into the sac, and thence along the Fallopian tube to the uterus. They cannot enter the body cavity, being shut off from it by the sac wall.

It seems impossible to arrive at any satisfactory conclusion concerning the function of this ovarian sac. In its more or less complete forms it evidently prevents or greatly diminishes the possibility of the shed ova passing into the general cavity of the abdomen, and consequently it precludes or lessens the chance of occurrence of abdominal pregnancy; at the same time it ensures the more rapid passage of the ova into the Fallopian tube. Under these circumstances it might be thought that a well-developed sac would be met with in all those animals in whose case it is necessary, for the preservation of the species, that all the mature ova should be fertilised. Such animals would arrange themselves in at least two classes—

1. Animals in which but few ova reach maturity.
2. Animals whose life is short and precarious.

There seems, however, to be no proof that this is the case. Undoubtedly the sac is present in its most complete form in rats and mice, both necessarily prolific animals, but it is present in its simplest form in the rabbit and cat, both of which animals produce many young at a birth, and bring them forth at frequent intervals. Again, the pouch is well developed in the elephant, an animal that gives birth to few offspring at long intervals. Indeed, the sac is present in a more or less well developed form throughout all the mammalian orders, and all the modifications of it, from the simplest to the most complete, are presented by one order, namely, the order of rodents.

It is interesting to note that this shutting off of a portion of the body cavity to form an ovarian sac is neither confined to mammals nor to vertebrates, but is also found in the invertebrate kingdom.

In the Chætopods the ovary is placed in the body cavity, and the ova are shed from its surface into that cavity, and escape from it by an oviduct. In leeches the ovary is enclosed in a sac which communicates with the exterior by an oviduct, but has no opening into the body cavity. If, as Balfour states,* the affinities of the leech are with the Chætopoda, this ovary must have become shut off from the body cavity in a secondary manner; and Bournet supposes the shutting off to have occurred at some time prior to the appearance of hæmoglobin in the blood fluid, for he finds the ovarian sac filled with a fluid similar to the blood, except that it contains no hæmoglobin.

Again, in some fishes the ova are merely cast off from the ovaries into the body cavity, from which they escape by abdominal pores (*Marsipobranchi*). In others an oviduct is present, the abdominal orifice of which lies close to the ovary (*Elasmobranchii* and *Ganoidei*). In many Teleostei the ovary is placed in a sac which lies in the abdominal cavity, and is attached to the abdominal wall by a mesovarium. The lower portion of the sac is lined by a different epithelium, and forms a duct leading to the exterior of the body It seems very probable, taking into consideration what occurs in mammals and leeches, that the sac-like teleostean ovary is a secondary formation, and is developed in some way from those cases in which there is an ovary and a Fallopian tube.

There appears, however, to be no evidence that such is the case and the development of the sac-like ovary has not, so far as I have been able to ascertain, been described.

---

* *Comp. Embryol.*, vol. i., p. 35.
† *Quar. Jour. Micro. Soc.*, No. xlv., 473.

Finally, I wish to thank Professor Young for many valuable suggestions, and for his kindness in allowing me to make use of several specimens, and my colleague, Dr. Paterson, for the drawing of the ovarian capsule of *Cynocephalus porcarius*.

EXPLANATION OF PLATE IV.

- *b.* Broad ligament.
- *c.* Cavity of ovarian sac.
- *f.* Fallopian tube.
- *f'.* Orifice of Fallopian tube.
- *o.* Ovary.
- *s.* Wall of ovarian sac.
- *r.* Uterus.

Fig. 1 represents the ovary, Fallopian tube, and simple peritoneal sac of a cat.

Fig. 2 represents the more completely-formed sac of *Cynocephalus porcarius*.

Fig. 3 represents the ovarian sac, &c., of the porcupine.

Fig. 4 represents a portion of the broad ligament, and the ovarian capsule covering the ovary of the racoon. The minute orifice of communication between the ovarian sac and the peritoneal cavity is also shown.

Fig. 5 represents a section of the ovary and ovarian sac of an adult mouse.

Fig. 6 represents a section of the ovarian sac, ovary, and termination of the Fallopian tube in a rat three days old.

Fig. 7 represents a section of the ovarian sac of an adult mouse, and shows the fimbriated end of the Fallopian tube projecting into the sac.

Plate IV

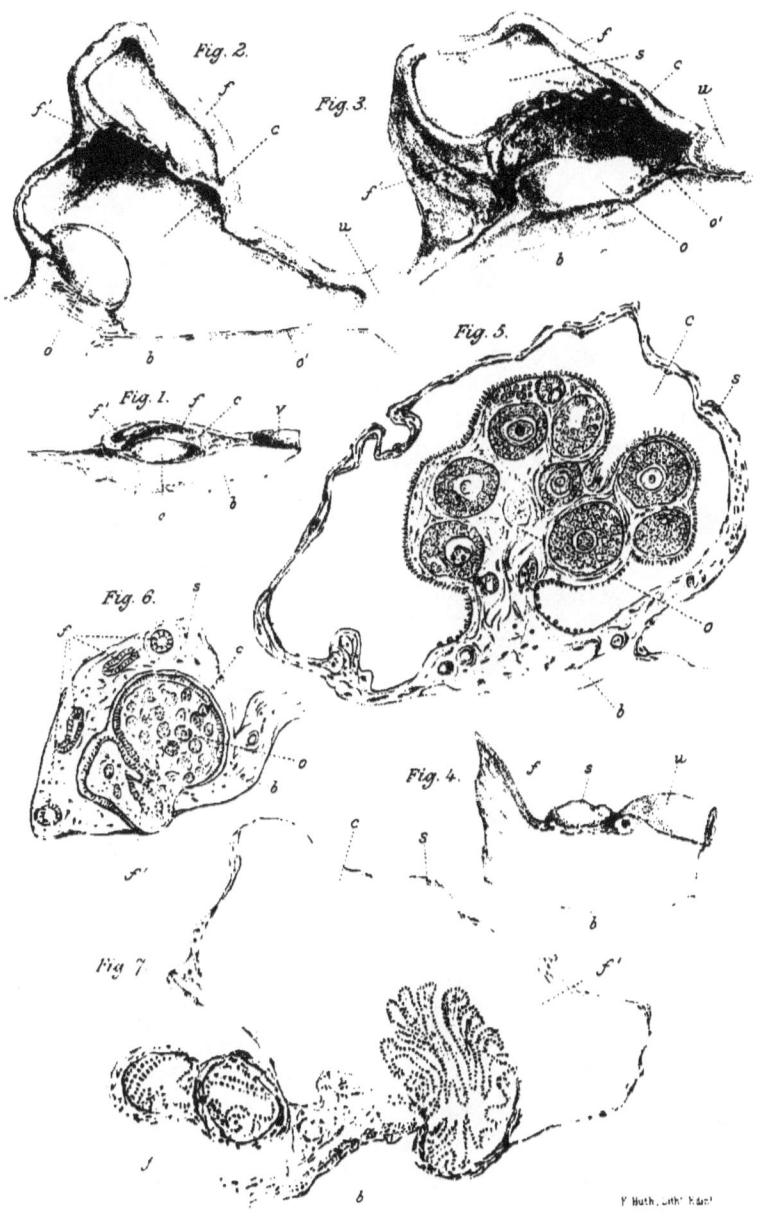

RELATIONS OF MAMMALIAN OVARY.

THE PUPAL STAGE OF CULEX.

AN INAUGURAL DISSERTATION FOR THE DEGREE OF PH.D. IN THE UNIVERSITY OF LEIPZIG.

By C. HERBERT HURST, *Lecturer in the Victoria University, Assistant Lecturer in Zoology in the Owens College.*

[WITH PLATE V.]

SHALLOW pools in most parts of Europe, and especially the smaller pools in woods, swarm in early spring with larvæ of Culex, hatched from eggs laid in floating masses by the impregnated females which have lived through the winter. These larvæ have been described by Swammerdam (1) and others, and most recently by Raschke (2).

After a few weeks the pupa escapes from the larval cuticle, and four days later the perfect insect flies free.

Though the pupal stage is the one of which I propose to give a fuller account than has yet appeared, it is necessary to the proper understanding of it that some account of the preceding and following states should also be given, and especially of the mode of life in each state.

The *larva* is an exceedingly active creature, swimming by a wriggling movement of the body, this being aided by a median fin-like series of setæ beneath the last segment. The head is provided with jaws and setæ, by means of which the solid food is collected and masticated. A pair of unjointed antennæ of considerable length arise from the sides of the head, and behind the base of each is a compound eye and an ocellus.

The head is moveably attached by a neck to the broad, rounded thorax. The abdomen is long and slender, and composed, apparently, of nine segments. The ninth segment bears four gill-plates surrounding the anus, and on its ventral surface, the median series of long setæ which serves as a propeller.

Respiration, according to Raschke, is performed in a threefold manner; by the gills just mentioned, by the rectum, and directly, the air being taken into the tracheæ by a conspicuous siphon projecting upwards from the eighth abdominal segment. The tracheæ of the abdomen serve not only as organs of respiration, but also by virtue of their great size as a float, keeping the larva when at rest at the surface of the water, with the hinder end upward and the end of the siphon touching the surface.

The alimentary canal is practically straight. The œsophagus is narrow. The stomach is wide, and extends from the anterior part of the thorax to the sixth abdominal segment. Its walls are very thick, and the epithelial cells very large, and in the thorax it has eight diverticula or pouches. From its hinder end the small intestine runs backwards to open into the wide rectum at the anterior end of the eighth abdominal segment, and this leads direct to the anus at the end of the body.

Five Malpighian cæca lie in the hinder segments of the abdomen, and open into the anterior end of the small intestine.

A pair of sac-like salivary glands lie at the sides of the stomach in the anterior part of the thorax, and their ducts, according to Raschke, unite, and have their opening " oben am Beginn des Œsophagus."

So far I have followed Raschke's account, except in ascribing a hydrostatic function to the "colossal" tracheal system, which is apparently much larger than would be necessary for respiratory purposes alone.

To Dr. Raschke's account I would add that, in addition to the head appendages, there appear during the larval period not less than eight other pairs of appendages beneath the larval cuticle. Of these six pairs are thoracic and two abdominal. The thoracic appendages are three pairs dorsal, the future pupal siphons, the wings and the halteres; and three pairs ventral, the future legs. The two abdominal pairs belong to the two last segments. Those of the eighth abdominal segment lie in the larval siphon, and form the fins of the pupa; the hindmost pair form the outer gonapophyses of the adult, which are accessory organs of copulation.

All these eight pairs alike arise as foldings of the epidermis ("hypodermis") outwards. All alike are completely hidden by the larval cuticle.

The antennæ, moreover, are much larger in an advanced larva than they appear to be. Their growth forwards being prevented by the unyielding cuticle, they grow backwards, and their basal portion is folded, and even "telescoped."

Towards the end of larval life the animal becomes sluggish; profound changes in its mouth-parts deprive it of the power of eating, and it floats with its siphon-stigma at the surface. Shortly the cuticle bursts in the thoracic region, along the mid-dorsal line; the pupal "horns" or siphons are protruded, the abdominal tracheæ appear to collapse, and the animal floats with the anterior end upwards, the new siphons coming to the surface. The old larval siphon, or rather its soft parts, are withdrawn from the cuticle and *invaginated into the eighth segment of the abdomen;* the intima of the abdominal and thoracic tracheal trunks breaks up into pieces, which in the abdomen correspond to body-segments. The body of the escaping pupa is gradually withdrawn from the larval cuticle, and the eighteen fragments of the old tracheal intima are drawn out of the body by nine pairs of stigmata, and cast off with the exuviæ. These nine pairs of stigmata are situated, one in the hinder part of the thorax, one in each of the first seven segments of the abdomen, and the ninth pair are united to form a single aperture, the old respiratory opening at the end of the larval siphon.

With the larval exuviæ are also cast off the cuticular portions of the jaws and antennæ, and all the hairs and spines with which the larval cuticle was beset.

The *pupa* which thus escapes differs from the larva very widely. It is a little under 1 cm. in length when fully extended. It consists of a bulky, laterally compressed mass made up of head and thorax with their appendages, and of a slender flexible abdomen, which when at rest is curved under the thorax. In a specimen measuring 9 mm., which is nearly the maximum size, the thorax measures 2·5 mm. and the abdomen 6·5 mm., but the thorax appears to be much longer on account of the wings which extend downwards and backwards from its sides.

The head lies below the thorax, and so adds nothing to the length of the animal. It is broad from side to side, short from back to front, while ventrally it is drawn out into a long process which extends backwards under the thorax as far as the anterior part of the abdomen, where it curves upwards. This process is made up of what are usually spoken of as the "mouth-parts," including labrum, epipharynx, one pair of mandibles, two pairs of maxillæ, and the hypopharynx. The second pair of maxillæ are fused together to form the labium.

In describing an animal which is coiled up so that head and tail almost meet, the terms "dorsal," "ventral," "anterior," and "posterior" are liable to be misleading. To avoid this as far as possible I shall apply the terms to those parts to which they would be respectively applicable in the fully-developed insect in the act of sucking blood, *i.e.*, I shall regard the general direction of the mouth-parts as downward, their distal ends as ventral, and I shall speak of the labrum as being in front of the mouth.

From the sides of the epicranial region the antennæ run outwards to the sides of the thorax, and then downwards, one beneath the anterior margin of each wing. The head and all its appendages are immoveable during pupal life.

The thorax is rounded, but somewhat compressed from side to side. From the sides of its summit arise the respiratory siphons, a pair of conspicuous organs whose position has led to the name "horns" being applied to them. The wings are nearly flat oblong plates, arising behind the bases of the siphons and extending downwards and backwards. Immediately behind them are the halteres, a pair of triangular plates enclosing the halteres of the future gnat. I have endeavoured to show the forms of these parts in Fig. 1.

The legs are almost completely hidden by the wings, but the femur, tibia, and first joint of the tarsus of the first leg, and the tibia and first joint of the tarsus of the second are visible (Figs. 1 and 2).

The abdomen is flattened dorso-ventrally, and when at rest is curved under the thorax. It is jointed and flexible, and forms with the pair of large flat fins borne by its eighth segment the only locomotor organ of the pupa, the wings and legs lying immoveable and even adhering to one another, though they are easily separated in specimens which have been kept in alcohol.

The pupa does not eat. It breathes air through the apertures at the ends of its siphons. It floats, thorax upward, by virtue of a large air cavity lying under the hinder part of the thorax and the anterior part of the abdomen. This cavity is bounded in front by the legs, at the sides by the wings, and below by the mouth-parts. It extends up at each side of the first segment of the abdomen, where it is covered by the halteres, and into this part of the cavity at each side opens a large stigma, held open by the fairly well-developed cuticular lining ("intima"), and guarded near its entrance by numerous spines. These two stigmata belong to the first abdominal segment, and put the air-cavity just described into direct communication with the tracheal system. As already mentioned, I regard this cavity and these stigmata as being mainly, if not exclusively, hydrostatic in function, serving not only to make the pupa float when at rest, but to make it float in a definite position, with the thorax upward and the apertures of the siphons at the surface of the water.

The pupa is sensitive to light, and immediately darts backwards when a shadow falls upon it suddenly. The movements, however, though very rapid, are devoid of anything like steering. The larva had to steer in its search for food, but the pupa has simply to get out of the way of danger, and the direction of its flight is of little importance, though, since the movement is always backward with reference to the pupa, it is chiefly downward with reference to the outer world.

A sudden very loud noise, or a very gentle tap upon the vessel containing the pupæ, causes those at the surface to dart downwards, but as slight sounds of various kinds produce no effect upon them, I conclude that the tremor of the surface of the water, and not the sound itself, was recognised by them. The setæ on the first segment of the abdomen are probably the organs by which this movement is felt.

As to the anatomy of the pupa, it is only necessary now to state that at the beginning of pupal life the internal arrangements are those of the larva; at the end of that period they are those of the imago.

At the beginning of the fifth day of pupal life, the cuticle splits along the mid-dorsal line of the thorax; the thorax of the imago protrudes, and the head, then the abdomen, and lastly the wings, legs, and proboscis are drawn out of the pupal cuticle, which is left floating in the water while the imago flies away.

With the cuticle are cast off nine pairs of fragments of tracheal intima, two pairs being drawn out through the thoracic stigmata, the others through the stigmata of the first seven segments of the abdomen.

These fragments differ as follows. The first pair are well developed, and have the spiral thickenings very well marked. They are continuous with the lining of the respiratory siphons, and formed during pupal life the connection between these organs and the tracheal system generally.

The first abdominal pair are not so well developed, but the spiral thickening is recognisable in them, and the terminal portion of each is better developed than the rest, and is beset internally with numerous small spines. It was through these that the tracheal system communicated during pupal life with the air-cavity beneath the thorax. The remaining fragments, *i.e.*, the hinder thoracic pair, and all the abdominal pairs except the first, are very thin and delicate, and were functionless during pupal life.

The cuticular lining of the anterior and posterior portions of the alimentary canal and the cuticle of the invaginated larval siphon are also shed, together with all setæ and the whole of the pupal siphons and fins. These three last alone involve the loss of portions of tissues other than cuticle.

The imago, with its long, slender body, wings, legs, and proboscis, hardly needs to be described. Like pupa and larva it breathes air, but now by more numerous stigmata, and unlike them it flies in the air. The larva fed upon solids; the pupa did not eat at all. The imago feeds upon fluids, and the female, at least, upon the hot blood of man and other mammals. The male is short-lived, and his food is said to consist of the sweet juices of flowers. To find the female and to impregnate her are the real objects of his short life. His antennæ are provided with long hairs, which A. M. Mayer (**3**) has shown to be sensitive to a particular sound when the head is turned towards the source from which it proceeds, and he has further shown that sound to correspond to the note emitted by the vocal organs which Landois (**4**) has described on the sides of the thorax of the female, beneath the halteres. The male, therefore, would appear to be endowed with a special and very largely developed pair of organs for detecting the whereabouts of the female.

The female imago, after impregnation, has to find a suitable place to lay her eggs, i.e., the surface of a stagnant pool; and having found it, has to lay the eggs at a suitable season, and this in the case of females escaping from the pupa late in the summer involves the necessity of living through the winter. Hence the mouth appendages are specially adapted for piercing the skin and extracting the blood of mammals, and this is stored in her capacious stomach. The structure of the mouth-parts in the two sexes has been described by Dimmock (5); but as he appears uncertain as to the injection of "saliva" into the wound, I shall add that a special apparatus developes during pupal life by which the saliva is discharged near the tip of the hypopharynx ("lingua").

Having now given a brief outline of the life-history of the gnat, I will proceed to describe the pupa and the changes which it undergoes in more detail. As, however, I have directed my attention almost exclusively to the more important organs of the body, and not to hairs and the like, I shall not make any distinction of species. What I have to record is probably applicable to all species alike.

DESCRIPTION OF THE PUPA OF CULEX.

The External Characters.

The *head* is broad from side to side; the epicranium has a well-marked median groove; the clypeus, broad above, is gradually narrowed below, and continued without any distinct line of demarcation into the labrum. At the sides are a pair of compound *eyes*, to be regarded rather as the rudiments of the eyes of the future gnat than as the visual organs of the pupa itself. Their form and size in the earliest stage are shown in Fig. 1. During pupal life they increase in size till they almost encircle the head. Corneal facets are never formed in the pupal cuticle, but beneath it the convex facets of the imaginal cornea are formed during pupal life.

Behind the compound eye, on each side of the head, is an *ocellus* with fully-developed lens, etc. In the youngest pupæ it is separated by a small interval from the compound eye (see Fig. 1); but the growth of the latter obliterates this interval, and the ocellus is in the older pupæ not readily distinguishable except in sections. The statement, found in systematic works, that the Tipulariæ are devoid of ocelli is, however, not strictly true; in Culex, at least, they are well developed, though, as they abut upon the compound eye, they are in the imago so inconspicuous that they may easily be overlooked.

In the *mouth-parts*, the labrum, epipharynx, mandible, maxillæ with their palps, labium and hypo- and epipharynx are present, though the two last can only be seen on dissection.

Of their mode of origin in the larva I as yet know nothing. At the time of escape of the pupa from the larval cuticle they are of the full size, which is considerably greater than in the adult. The form of most of these parts is shown in Figs. 1 and 2. That I may not have to refer to these parts again, I will at once say that the chief changes which occur in them during pupal life are:— (1) The development of a cuticle within the pupal cuticle, and this, in the case of the labium (fused second maxillæ), is covered with scales closely resembling those found in Lepidoptera; (2) a considerable shrinking; (3) in the male only, atrophy of the mandibles, which in a young pupa are as large as in the female, but in the adult are not recognisable.

The *antennæ*, which were folded and telescoped at their bases in the larva, are in the pupa laid upon the sides of the thorax, as seen in Fig. 2. Their hinder (distal) extremities are hidden by the wings. The swollen basal joint of the antenna of the imago is hardly recognisable on the surface, although it is already a conspicuous object in sections of the youngest pupæ, and even in the larval state. I shall describe it with the other sense organs. The shaft of the antenna is segmented, but the external segmentation loses its correspondence with the segmentation of the developing antenna within it early in pupal life.

The *thorax* is large and rounded, but somewhat compressed from side to side. Mid-dorsally the cuticle of the prothorax is marked by fine transverse corrugations, and this is the part which ruptures to allow the imago to escape. A pair of branched setæ arise from the dorsal region of the hinder part of the thorax.

The *respiratory siphons* (AT, Fig. 1) are nearly cylindrical, narrowed at their bases and curved forwards to be attached by flexible membranes to slight prominences on the sides of the prothorax. Above they are obliquely truncate and open, and the margin is slightly notched on the inner side. The outer surface is marked so as to resemble imbricated scales, each with a minute spine at its apex. The cavity of the siphon communicates directly with that of a tracheal trunk at its base. Palmén (6) says that after a "close investigation" he has found that there is no opening. The tone of assurance in which he contradicts all previous observers led me to put the question to the test. I removed the side wall of the thorax, with some of the underlying muscles and tracheæ, from a specimen preserved in alcohol. I drew out the alcohol from the cavity of the siphon by means of blotting-paper, and then touched the tip with a minute drop of glycerin. I watched the effect under the microscope, and saw the glycerin force its way into the siphon, driving the air before it into the tracheæ. Palmén, moreover, says the organs are gills! Each is a thick chitinous tube, the cavity guarded by numerous hooked spines, the walls consisting of hardly anything but the chitinous cuticle, the epidermis ("hypodermis") between its two layers being barely recognisable on account of its thinness. The "tracheal gills" on which Palmén lays much stress have absolutely no existence.

The *wings* of the pupa, that is the organs within which the wings of the imago are developing, are a pair of oblong plates about $2\frac{1}{2}$ mm. in length. They are closely applied to the sides of the hinder part of the thorax, and directed downwards and backwards. They are immoveable.

The *halteres* are a pair of elongated triangular plates lying along the dorsal and hinder border of the wings.

All these three pairs of dorsal appendages arise within the larva in the same way, and their bases or points of attachment all lie in the same horizontal plane. Each is at first (in the larva) a fold of the epidermis; each acquires a cuticular covering (like all other parts of the body), and the first pair become rolled up to form tubes, the respiratory siphons, while the other two remain flat plates.

The three pairs of ventral appendages of the thorax, or *legs*, are long cylindrical bodies folded upon themselves, and lying beneath the thorax and between the wings. The same segmentation into femur, tibia, etc., is recognisable as in the adult gnat, but the segments are more nearly equal in the pupa, and the joints of the developing and shrinking legs of the future imago soon lose their correspondence with those of the pupal cuticle enclosing them. They arise in the larva, like other appendages, as folds of epidermis enclosing mesoblastic tissues.

The *abdomen* is dorso-ventrally compressed and exceedingly flexible dorso-ventrally, though not from side to side. It is the only part of the pupa in which the segmentation of the body is readily recognisable, and as I shall very frequently have to refer to the various segments by number, *I shall use the terms "first segment," etc., to signify "first segment of the abdomen,"* etc.

Nine segments are readily recognised in the abdomen, and the last one, though it is probably composed of no less than three condensed and highly modified segments, I shall call simply "ninth segment."

Each abdominal segment has a chitinous tergum and sternum, and setæ are distributed sparingly over them, being almost confined to the hinder parts of the terga. The terga and sterna of successive segments are united by soft arthrodial membranes.

Of the *setæ*, only one pair need special mention. These are placed on the hinder part of the first segment, the base of each being a triangular plate attached by one angle to a soft membrane, and the distal side of the plate is divided into a number of bars which, by repeated division or branching, give rise to about one hundred setæ all lying in one plane parallel to the median plane of the body. Each seta bears a few fine hairs. When at rest, the pupa floats with the tips of these setæ, and the tips of the respiratory siphons, at the surface of the water, and these setæ probably assist in maintaining the equilibrium of the animal in this position, as well as serving as sensory organs by means of which any disturbance of the surface is felt.

The eighth segment bears a pair of large *fins*, thin oval plates about 1·2 mm. in length, attached by the narrow end beneath the tergum behind. Each is stiffened by a midrib which projects beyond the hinder border of the fin as a spine. (Fig. 2.)

Beneath the fins and behind the eighth segment is the "*ninth segment*" with its appendages. Though this region is probably made up of more than one segment, its composite nature is not easy to recognise, as the plates supposed in other insects to represent the terga and sterna of tenth and eleventh segments [see, for instance, Huxley (7) and Miall and Denny (8)] are not developed in the young pupa, nor, indeed, is there in any stage any such development of the pupal cuticle, though plates developed within as parts of the imaginal cuticle may perhaps represent some of these parts.

The appendages of the "ninth segment" of the pupa are a pair of blunt processes arising below and in front of the anus, and directed backwards below the fins. They are much larger in the male than in the female. A pair of appendages are already recognisable in this region in sections of the larva, and I think even two pairs, but this portion of the larva is particularly difficult to cut, and I am not yet certain as to the hinder of the two pairs. Of the existence of one pair I have no doubt.

The Digestive System.

The *alimentary canal* of the pupa runs almost direct from end to end of the body, the only convolution occurring in the region of the intestine.

In the youngest pupa the condition is practically that of the larva. (*See* Raschke, *op. cit.*) The narrow œsophagus projects slightly into the stomach. The stomach extends from the anterior part of the thorax to the end of the fifth segment (abdominal): it is very wide, and the eight diverticula found in the thoracic region of the larva are still present. The walls are very thick, and the cells of its epithelium large.

The stomach opens behind into the intestine, which is slightly coiled and opens into the wide rectum, which ends at the hinder end of the abdomen. The epithelium of the rectum consists of very large cells, and is thrown into longitudinal folds.

The *salivary glands* are unbranched sac-like glands in the anterior part of the thorax at the sides of the alimentary canal. Their ducts unite beneath the sub-œsophageal ganglia, and from this point the single median duct runs forwards to open in the floor of the mouth.

The five *Malpighian cæca* open into the anterior end of the intestine. They are nearly cylindrical bodies of an intense white colour; their closed ends lie in the seventh or eighth segment, and measure about 0·13 mm. in diameter. They run forwards, almost straight, to near the anterior end of the fourth segment, and then backwards to their point of opening into the intestine immediately behind the constriction dividing the latter from the stomach. The diameter of each near its opening is about 0·03 mm. Each cæcum is made up of two rows of cells alternating more or less regularly on the two sides, the narrow lumen taking a zigzag course between them. The individual cells are very large, the long diameter of each being the diameter of the organ itself. The nucleus is large and transparent, but the rest of the cell contains a large quantity of a granular deposit which gives the organs their intense white colour. During pupal life I have noted no important changes in these organs.

Such is the structure of the alimentary canal and its appendages at the commencement of pupal life—a structure adapted to the life of the larva, but not to that of the imago, and the changes which it undergoes during the pupal period are so great that at the end of that period no part of the whole canal corresponds in structure to the above description.

The most striking change is the reduction in thickness of the epithelium which occurs throughout, but which is perhaps best shown in the stomach. Four stages of the change are shown in Figs. 3, 4, 5, and 6, which are drawn from the epithelium of the hinder part of the stomach. The beginning of the change has already occurred before the pupa leaves the larval exuviæ, but the first stage here shown (Fig. 3) is from a young pupa. At the base of each of the large epithelial cells may be seen one or two nuclei; a little later the protoplasm of the cell divides into a small portion around the new nuclei, and a much larger portion, which rapidly undergoes degeneration and, separating from the basal layer (the new epithelium), is apparently digested. The outer surface of the stomach is covered by an exceedingly thin layer in which I could not make out any structure, but which is presumably muscular, and is at first folded longitudinally (Figs. 4 and 5), but afterwards becomes even, the new cells at the same time becoming flattened (Fig. 6).

A similar change occurs in the intestine. The epithelium divides into a thin outer and a thick inner layer. The latter becomes loosened, breaks up, and appears to be digested.

In the rectum more complex changes occur, though here also the superficial portion of the epithelium is thrown off, but it breaks up later and more slowly than elsewhere; in fact, the disintegration and digestion appears to commence in the anterior part of the stomach, and progress gradually backwards. Before the epithelium shows any sign of disintegration in this region, the rectum becomes differentiated into two parts: an anterior very wide part, the "rectal pouch," and a narrower hinder portion, to which alone I shall apply the term "rectum" from this stage onwards. The wall of the rectal pouch rises up into four very large and prominent papillæ, the "rectal glands": one ventral at the anterior end of the pouch, just below the opening of the intestine into it; one dorsal and posterior, and two lateral, intermediate in position between the other two. Nerves and tracheæ push their way into the axis of each papilla. The epithelium of the papillæ undergoes division into two layers as elsewhere, but the distal layer, which is ultimately shed, is very thin, and the basal or permanent epithelium consists of very large columnar cells, while the opposite is the case everywhere else, and especially in the rectum, where the permanent epithelium is so thin that I had difficulty in detecting it, and Chun (9) states that it is absent in *Musca* and other insects.

Besides this shedding of epithelium, changes of form occur in various parts of the alimentary canal.

The anterior part of the œsophagus expands, especially in the female, and acquires a thick chitinous lining. In cross section it becomes triangular, and the sides and roof, all of which are convex inwards, are supplied with muscles arising from the walls of the head, which by their contraction increase the size of this cavity, and serve to produce the sucking action by which the imago draws the blood of its victims up through its proboscis. This apparatus is not well-developed in the male.

The posterior part of the *œsophagus* gives off ventrally a large diverticulum ("crop"), which runs backwards under the stomach as far as the hinder end of the thorax, its walls developing numerous small sacculations along its two sides. Sometimes a forwardlydirected diverticulum of this crop is found arising from its ventral wall.

The cavity of the *stomach* becomes wider, while the part behind it becomes narrower, with the exception of the rectal pouch.

The *salivary glands*, which at the beginning of pupal life were a pair of hollow unbranched club-shaped organs lying at the sides of the alimentary canal in the anterior part of the thorax, become during pupal life divided into about four branches, and the cavity almost disappears, and acquires a pretty thick chitinous lining. The *ducts* run downwards to the neck, which they traverse at the sides of the nerve cords. Just below the hinder border of the sub-œsophageal ganglia they unite to form a median duct, which runs forwards to open into a pit at the base of the hypopharynx. This pit becomes deeper during pupal life, and acquires a very thick chitinous lining. From it a deep groove, also *very* strongly chitinised, runs downwards along the middle line of the anterior surface of the hypopharynx to its extremity. This is true of both male and female; but the hypopharynx of the male is inseparable from the labium.

The Circulatory System.

The *heart* lies in the abdomen in a median space between the extensor muscles and close beneath the dorsal wall of the body. It arises at the anterior end of the eighth segment, and ends suddenly at the anterior end of the first segment, giving off the aorta from the ventral border of its anterior end. From its sides "alæ cordis" run outwards beneath the extensor muscles and between the main tracheal trunks and the stomach, to be attached to the "peritoneal" covering of the tracheal trunks, or to the outer layers of the wall of the stomach. Each "ala" consists of a dorsal and a ventral lamina, the two running together some distance from the heart. The space between them has been called "pericardium": it contains the "pericardial cells," and communicates freely with the body cavity by the spaces between the alæ. The ventral lamina of each is continuous with the corresponding lamina of the other side of the body, and all the ventral laminæ together thus form an imperfect "pericardial septum" (Graber). The dorsal laminæ are attached to the sides of the heart near the dorsal surface, their fibres taking a longitudinal direction on the heart, and forming its outermost layer. The heart is further bound by fibrous strands to the dorsal body wall.

Graber (10) appears to believe that the septum is invariably attached to the outer wall of the abdomen, dividing the cavity of the abdomen into two cavities, a small dorsal "pericardium" containing only the heart and pericardial cells, and a large ventral cavity containing all the other organs of the abdomen. His figure of Acridium is reproduced in the most popular text-book (Claus), and his view that this arrangement is universal, and that the "septum" serves as a pump driving blood from the large abdominal cavity to the pericardium, is reproduced in other text-books, in such form as to lead to the belief that the arrangement is the same in all insects. Whatever may be the case in other insects, this view is certainly not applicable to Culex. Here the "septum" does not extend to the body wall, and if a "pericardium," in Graber's sense of the term, exist at all, the extensor muscles and the main tracheal trunks lie in it, and the septum cannot, judging from its anatomical relations, have the function ascribed to it.

The heart itself is a more or less cylindrical tube, about 0·06 mm. in diameter. Its hinder end at the anterior limit of the eighth segment is open, but I am unable to give an account of any valvular apparatus which may be present here. There is no sharp division into chambers either by constrictions or by valves. In the first segment a pair of valved ostia opens *backwards;* in segments III to VII paired ostia are present, their margins being turned in and directed *forwards* to form the valves. I have not detected any aperture or valve in the second segment. The ostia are small paired slits in the sides of the heart, and between the two laminæ of the alæ, putting the cavity of the heart in communication with the "pericardial" cavity. The infolded margins of the slits serve as valves in two ways; first, they prevent the blood from flowing out through the ostia; and, second, they prevent the blood within the heart from flowing backwards during systole.

Of the histology of the heart I would speak with the greatest caution. Graber (*op. cit.*) has made the subject his own, and has applied very special methods to the investigation. My object has been rather to record the anatomical structure and the development of the pupa, and I simply note histological results incidentally, referring those who wish to learn the histological structure of insect hearts to the classical work just mentioned.

Without the application of special methods, I have recognised three layers in the wall of the heart.

The inmost layer, or *endocardium*, is an exceedingly thin layer of flat cells. Their nuclei are conspicuous objects, occurring with striking regularity in pairs, four pairs to each segment of the abdomen, and a similar but smaller nucleus is to be seen in each flap of each valve, from which I conclude that this endocardium extends also to the valves. Whether the other layers also extend into the valves or not, I cannot say with certainty.

The *middle layer* consists of encircling fibres, slightly oblique in direction, and probably muscular.

The *outer layer* is also fibrous, its fibres being on the whole longitudinal in direction, but they curve outwards to be continuous with the fibres of the dorsal laminæ of the alæ.

Between the laminæ of the alæ cordis, that is in the pericardial cavity, are large ovoid masses of brown cells, the "*pericardial cells.*" Of these masses there are two pairs near the anterior, and two near the posterior end of each segment of the abdomen; but the number increases towards the end of the pupal stage, and still further in the imago, by the division of some of them into two or more masses. The protoplasm of these cells is extraordinarily spongy, and contains numerous granules, which stain deeply with borax carmine. The nuclei vary in number from three or four to ten in each mass, though the boundaries of so many cells cannot be made out. The cells appear to be undergoing division very slowly. The excretory function of these cells has recently been shown by Kowalevsky (12).

The *aorta* runs from the ventral border of the anterior end of the heart forwards above the stomach and œsophagus to the head, where it ends, the end being open. In transverse sections of the thorax, the aorta is seen as a laterally compressed tube. I have not seen any branches given off from it.

THE RESPIRATORY SYSTEM.

Culex, as already mentioned, breathes air in all three states— larva, pupa, and imago—and also breathes it directly, but the air is taken in at different apertures in the three states. The larva, according to Raschke (*op. cit.*), breathes also by gills and by the rectum.

I have already described the respiratory siphons of the pupa, and given evidence to show that they really do lead directly into the tracheæ, in spite of Palmén's contention to the contrary.

From the base of each siphon, tracheæ run to various parts of the body and head. Amongst these may be mentioned specially one transverse trunk running across the thorax between the alimentary canal and the nerve-chain, and putting the two siphons in direct communication with each other; and a pair of longitudinal trunks running backwards to the hinder end of the body, and giving off branches to the various organs, and also a trunk to each of the stigmata. As already mentioned, these stigmata are present in the hinder region of the thorax, and in each of the first seven segments of the abdomen; but the stigmata, except the first abdominal pair, are closed, and the pupal intima of the tracheæ connecting them with the main trunks is thin and collapsed. The widely open stigmata of the first segment, with their spines and their probable function, I have already commented upon; but while insisting on the importance of the hydrostatic function of the tracheal system in both larva and pupa, I would again say that I do not consider this a sufficient ground for the view that the hydrostatic function is the primitive one. In Culex larva and pupa, it is important only inasmuch as it subserves respiration by bringing the animal to the surface and maintaining it there in the only position in which air can be breathed directly.

The cuticular lining ("intima") of the chief trunks and their branches is well developed even at the commencement of pupal life, and has the usual spiral thickening. The trunks connecting the stigmata with the main trunks are the only ones that undergo any marked change during the pupal condition. These widen around their separated and collapsed intima, and a new and strongly thickened intima is formed. In the main trunks no new intima is formed, and when the imago escapes from the pupal cuticle no portion of the intima is shed from any part of the system which has been functional during pupal life, excepting the portions connecting the siphons and the first abdominal stigmata with the main trunks. These fragments are, in the case of the siphons, well developed, and have a fully-developed spiral thickening. The portions connected with the first abdominal stigmata, though better developed than the portions connected with the other stigmata of the abdomen, have the spiral thickening only slightly developed. The terminal portion is beset with very numerous small spines.

The fate of the invaginated portion of the larval siphon is interesting. The whole of the tissues composing it break up and undergo complete absorption, so that no trace of it is discoverable in the advanced pupa.

Before dismissing the respiratory system, I will again state that the pupa breathes air only, and breathes it through the open stigmatic horns or siphons alone. The *tracheal gills* of Palmén have no existence.

The Muscular System.

Concerning this I have nothing new to communicate. The muscles of the pupa are those of the imago. All the chief ones are present in the young pupa, but they increase greatly in size, and this is especially true of the thoracic muscles.

The Nervous System.

The *nervous system* is particularly interesting. Within the short space of four days, certain ganglia increase enormously in size by the addition of cells apparently derived directly from the epidermis; and other ganglia, already well developed and functional, shift bodily from their original positions, and in some cases fuse with ganglia originally remote from them.

Raschke (*op. cit.*) says that in the larva each of the first eight segments (of the abdomen) has a pair of ganglia, and this statement is certainly true of all the larvæ I have examined, and yet a pupa which I killed when only half escaped from the larval cuticle had already four pairs in the thorax, and none in the first segment of the abdomen. During pupal life these four ganglionic masses fuse into one compact mass, though its composite nature is always recognisable in sections.

The ganglia of the eighth segment (of the abdomen) at the beginning of pupal life occupy their typical position in the anterior part of the segment, and are connected with the ganglia of the seventh segment by connectives (or "commissures") nearly equal in length to the seventh segment. During the first two days of pupal life these connectives vanish completely, and the ganglia migrate to the anterior part of the seventh segment to fuse with the ganglia of that segment. As with other composite ganglionic masses, the composite nature of the ganglionic mass so formed is easily recognised in sections, especially horizontal sections, even in the imago. In the female the process goes a stage further. A pupa almost ready to burst and give exit to the imago has still the arrangement just described, but the imago, killed immediately after its escape, is found to have no ganglia in the seventh or eighth segment, but in the sixth segment are two masses; the first, the pair properly belonging to the segment, lying at its anterior end; the other, the double ganglionic mass formed by the fusion of the seventh and eighth ganglia, lying at the hinder end of the segment.

In the *male* imago, however, the arrangement is the same as in the advanced pupa.

A detailed description of the ganglia of the head and the changes they undergo during pupal life would take me too far. The most striking change is the very great increase in size which these ganglia undergo, and the most interesting point is the way in which this increase is brought about. The epidermal ("hypodermal") cells, especially those near the borders of the eyes, proliferate freely, and the cells budded off from their inner surfaces migrate inwards and form the new cells of the ganglia. By this process the ganglia, which at the beginning of pupal life were comparatively inconspicuous, grow till they almost fill the head, and there are places in the advanced pupa where the ganglia and the epidermis appear to be continuous.

The *sense-organs* of the pupa itself are not of special interest, that is, the organs which serve during pupal life as sense-organs. The *setæ* have already been mentioned. The *ocelli* are those of the larva, but they persist to the adult condition, the chief change which they undergo being the development of an exceedingly dense pigment. I have already referred to the common but erroneous statement that the imago is devoid of ocelli.

The *compound eyes* belong properly to the imago, not to the pupa, though they are probably sensitive to light in the pupal condition. At first they are small (see Fig. 1) and devoid of corneal facets, but they grow till they occupy the greater part of the surface of the head.

So much has of late been written upon the eyes of insects, that one should hesitate to add to that literature without having made very special study of the organs in question. Still, one of the most remarkable papers of the day (Patten, **11**) has attracted so much attention, and is so strongly opposed to the views of previous observers, that the little I have already observed may be of interest.

Each eye is made up of a very large number of "elements." Growth of the eye consists in the addition of new "elements" at its edge. Each new element is formed directly from the previously unmodified epidermis at the margin of the eye, and each arises independently of the rest of the eye, as a separate invagination of the epidermis. The cells, four in number, around the margin of each invagination, persist as the "nuclei of Semper," "corneal hypodermis," "corneal epidermis," "cellules cristallines," "cellules de Semper," "refractive globules" or "spherules." The invaginated portion gives rise to all the other parts lying outside the limiting membrane, with the possible exception of the pigment cells. The elements are at first devoid of pigment.

The details of the development I have not yet worked out, and I think it best to reserve further description for a future paper.

Other sense-organs developed during pupal life are *antennæ*. Antennæ are, it is true, present already in the larva, but they have no resemblance to those of the imago, and they are functionless during pupal life.

The epidermis round the base of each antenna of the larva grows rapidly, and as it is prevented, by the rigid and unyielding cuticle of the shaft of the antenna, from growing forwards, it grows backwards, and becomes "telescoped" and much folded, and sections through the larva show that the differentiation of the epidermis of the different parts has already begun.

When the pupa escapes from the larval cuticle, much of the folding is undone, but a portion of the telescoping persists at the base of the organ, and this part gives rise to the large hemispherical basal joint of the antenna of the imago.

This remarkable organ was described in the imago thirty-five years ago by Johnston (**13**), but very imperfectly. Externally it is not conspicuous in the pupa, though it is just recognisable. During pupal life its parts undergo considerable change, and these will be best understood if I describe the adult structure first.

In the imago the antennæ differ markedly in the two sexes. In the female the shaft is longer than in the male, and the hairs with which it is beset are less numerous and very much smaller. In both sexes the basal joint is enlarged, and forms a nearly hemispherical cup, with small cavity and very thick walls, covered and lined with chitin. The shaft of the antenna arises from the centre of the cup, and the chitinous floor of the cup is strengthened by a series of radial thickenings. In the female the edge of the cup is turned in, so that the aperture of the cup is narrower than the cavity immediately below. The structure in the male is really an exaggeration of this; the edge is folded in so completely that it unites with the floor, and the walls of the cavity of the cup of the female thus come to be represented by a concave double disc, the two laminæ of which are closely united, and the space between them, the equivalent of the cavity of the cup in the female, is here obsolete. The attachment of the shaft to the floor of the cup appears to be rigid, and the organ would appear to be adapted for the perception of sound-waves coming in the direction of the axis of the shaft alone.

A section taken along the axis of the organ shows the following structures: A layer of flattened epidermal cells, next to the cuticle of the outer wall; then a layer of cells I shall call "ganglionic," thickest at the base of the cup, and continuous with the antennary lobe of the "brain." Between this layer and the inner wall of the cup is a double (perhaps treble) layer of long narrow rod-like cells, at right angles to the surface, that is, radiating from the centre of the cup.

These structures form a thick ring round the cup, perforated at the base by the nerve supplying the shaft of the antenna.

The basal joint is supplied by an enormously large nerve arising from the ventral portion of the supraœsophageal ganglion at the side of the œsophagus. This nerve is broader than the abdominal double nerve cord, and is independent of the nerve supplying the shaft of the antenna, which lies ventral to it. The nerve, after entering the organ, divides, one layer penetrating the "ganglionic" layer; another runs between the ganglionic layer and the layer of rods, and a third on the inner surface of this layer, supplying certain small rounded cells lying between this layer and the base of the shaft.

All the cellular layers of this organ are epidermal in origin, but the layer which I have called ganglionic stands, during the later part of pupal life, in direct continuity with the superficial layer of cells of the "brain," and this layer in turn is continuous with the deep layer of the epidermis of the head immediately behind the base of the antenna. Whether the continuity of the ganglionic layer of the organ with the brain is due to identity of origin, both being budded off from overlying epidermal cells, or to migration of cells from the brain into this organ, is difficult to determine, but the cells of the cerebral lobes ("hemispheres") are larger than those of the cup-like organ, while the latter resemble the cells of the inner optic lobes very closely in size and in mode of staining.

This organ is already a conspicuous object in sections of the larva, more conspicuous indeed than the "brain," but the differentiation of the layers is only completed late in pupal life.

The Reproductive System.

I. The *male* generative organs of the adult consist of testes, vasa deferentia, "prostatic glands," copulatory organ with a common pouch at its base, and two pairs of gonapophyses. Of these last the outer ones are a large pair of forceps for holding the female. Both pairs are probably segmental appendages, and I have already spoken of their origin in the larva.

The testes are a pair of cylindrical bodies already present in the larva at the sides of the intestine in the sixth segment. They are chambered, and the spermatic elements in the hinder chambers are more advanced than those in the anterior chambers. The length of each testis is that of the segment in which they lie.

The vas deferens of each side is a direct continuation of the wall of the testis, and is a very narrow tube running backwards, quite distinct from its fellow of the opposite side, but the two are closely bound together in their hinder parts, and they open behind into the common pouch.

The prostatic glands are a pair of elongated glandular tubes, apparently simple, but seen in sections to be double, though the cavities communicate behind before opening into the common pouch.

The common pouch is a dilatation of the ejaculatory duct at the base of the copulatory organ, and the latter is perhaps derived from a pair of abdominal appendages.

The hinder parts of the vasa deferentia appear to be developed as a forward outgrowth of the ventral wall of the common pouch, and the prostatic glands are lateral outgrowths of the same. The hinder part of each vas deferens is in some Culicidæ expanded to form a vesicula seminalis of considerable size, but this is not the case in Culex.

II. The *female* generative organs are a pair of ovaries, oviducts uniting behind to form a median oviduct, a median copulatory pouch and three spermathecæ opening into the last.

The ovaries correspond in size and position with the testes.

The median oviduct is formed by invagination in the region which I take to be the ninth sternum, while the anus opens at the posterior end of what I take to be the eleventh segment, so that there is no common cloaca. This invagination is already far advanced at the beginning of pupal life (Fig. 7), and during this period it grows forwards, keeping pace with the forward shifting of the last pair of ganglia, and at all stages lying just behind it, till the final ecdysis, when the rapid shifting of the ganglia leaves it behind. Its anterior end is, in the adult, near the anterior end of the seventh segment.

In the youngest pupæ three flattened invaginations, the future spermathecæ, lie upon the dorsal wall of this median oviduct. During the pupal period the anterior end of each becomes spherical and acquires a strong chitinous lining. The anterior ends of these organs remain stationary in the eighth segment throughout.

The bursa copulatrix is a dorsal outgrowth of the invagination which gives rise to the median oviduct, and is a small pouch lying just behind and above the genital aperture.

I am painfully conscious of the fact that the foregoing account of this interesting pupa is far from complete, but the pressure of other work prevents my adding anything considerable to it at present. As soon as I have time to do so, I intend to work out the details of the development of the eye, but fear it will not be possible before next summer.

In conclusion, I would express my best thanks, *first*, to the Council of the Owens College, through whose generosity I have been able to leave my work in Manchester; and, *secondly*, to my honoured teacher, Herr Geheimrath Professor Dr. Leuckart, to whom I am indebted for many kind hints, especially as to the literature of the subject, and also for the loan of very numerous books and papers.

I subjoin a list of the books and papers to which I have referred in the foregoing dissertation :—

1. SWAMMERDAM.—"Bibel der Natur."
2. RASCHKE.—"Die larve von Culex nemorosus." Berlin, 1887.
3. A. M. MAYER.—"Researches in Acoustics," "American Journal of Science," 1874.
4. LANDOIS.—"Die Ton- und Stimm-Apparate der Insecten," "Zeitsch. f. wiss. Zool.," xvii., 1867.
5. DIMMOCK.—"Mouth-parts of some Diptera." Boston, 1881.
6. PALMÉN.—"Zur Morphologie des Tracheensystems." Helsingfors, 1877.
7. HUXLEY.—"Anatomy of Invertebrated Animals."
8. MIALL and DENNY.—"The Life-history and Structure of the Cockroach." London, 1886.
9. CHUN.—"Bau, etc., der Rectal-drüsen bei den Insekten." Frankfurt a/M., 1875.
10. GRABER.—"Ueber den propulsatorischen Apparat der Insekten," "Arch. f. Mikr. Anat.," ix., 1873.
11. PATTEN.—"The Eyes of Molluscs and Arthropods," "Mitth. aus d. zool. Stat. zu Neapel," 1886.
12. KOWALEVSKY.—"Ein Beitrag zur Kenntnis der Excretions-Organe," "Biolog. Central-blatt," ix., 1889.
13. JOHNSTON.—"Auditory Apparatus of the Culex Mosquito," "Journ. Microscopical Science" (old series), vol. iii., 1855.

DESCRIPTION OF PLATE V.,

Illustrating Mr. Hurst's Paper on the Pupa of Culex.

Fig. 1. Side view of the male pupa ($\times$ 10).

Fig. 2. Ventral view of the female pupa partially extended ($\times$ 10).

Fig. 3 to 6. Successive stages in the metamorphosis of the epithelium of the hinder part of the stomach ($\times$ 225).

Fig. 7. Sagittal section of a very young female pupa ($\times$ 50).

Ant. Antenna. *Ao.* Aorta. *At.* Respiratory siphon. *B.* Buccal chamber. *CG.* Cerebral ganglion. *D.* Gastric pouch. *F.* Fin. *Fe* 1. Femur of the first leg. *G.* Ganglia. *Gn.* Outgrowth of "ninth segment," within which the gonapophyses develop. *Hr.* Halter. *H.* Head. *Ht.* Heart. *In.* Intestine. *Lb.* Labium. *Lbr., Lr.* Labrum. *M.* Malpighian tube. *M.Ap.* Its opening into the intestine. *MS.* Mesosternum. *Mt.* Metasternum. *Mx.* Maxilla (first). *Mxp.* Its palp. *N.C.* Nerve commissures and ventral cord. *Oc.* Ocellus. *Od.* Median oviduct. *Op.* Compound eye. *P.* Prosternum. *R.* Rectum. *S.* Aperture of salivary duct. *SD.* Salivary duct. *SG.* Suboesophageal ganglion. *Si.* Larval respiratory siphon introverted into eighth segment. *Sp.* Spermatheca. *St.* Stomach. *Ta* 1, *Ta* 2. Proximal joints of tarsi. *Ti* 1, *Ti* 2, *Ti* 3. Tibiæ. *Tr.* Trachea. *W.* Wing.

I., II., III., etc. First to eighth segments of abdomen.

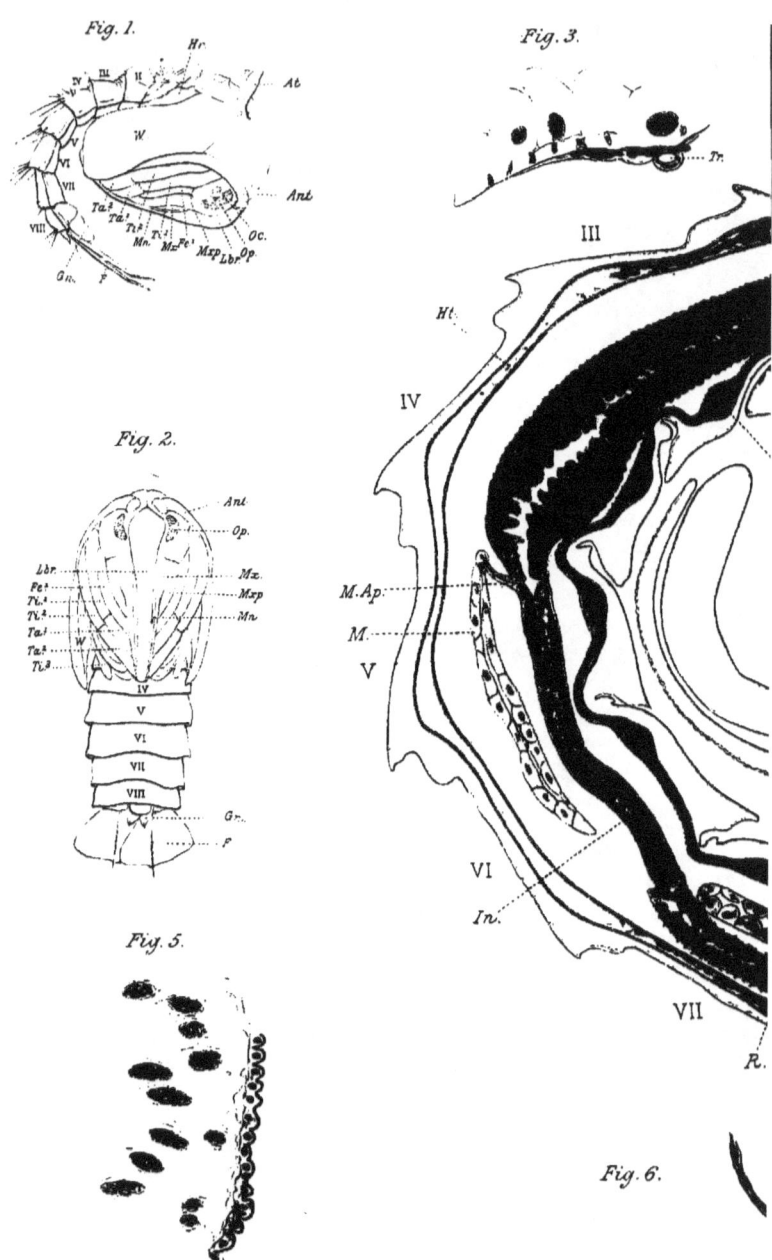

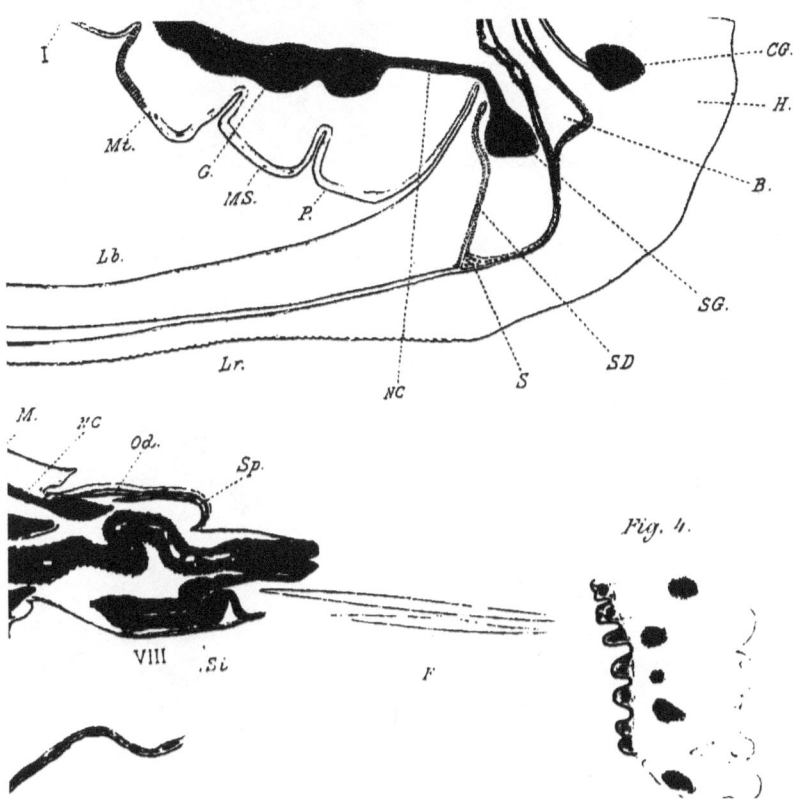

OBSERVATIONS ON THE STRUCTURE AND DISTRIBUTION OF STRIPED AND UNSTRIPED MUSCLE IN THE ANIMAL KINGDOM, AND A THEORY OF MUSCULAR CONTRACTION.

By C. F. MARSHALL, M.Sc., *Platt Physiological Scholar in the Owens College, Manchester.*

[WITH PLATE VI.]

STRIPED muscle has long been known to occur widely distributed in the animal kingdom, but the details of the structure of the striped muscle-cell have been the subject of much controversy. Various descriptions have been given, widely differing from one another, and none of them affording a satisfactory basis of comparison with other cells. The demonstration of an intracellular network in the muscle-fibre by several recent observers appears to afford the most rational clue to its structure, for it not only explains all the appearances seen in the muscle-fibre, including those seen with polarised light in the living fibre by Brücke, but it also renders possible a comparison with other cells, and shows that a muscle-fibre is to be regarded as of essentially the same structure as an ordinary cell, and must not be considered as an enigmatical structure, the details of which do not correspond to those of any other cell in the animal economy.

It is necessary first to examine the descriptions of the several observers who have described a network in the striped muscle-fibre, and to consider the interpretation that they have put upon it. In the following account I have only referred to those observers who have described some form of network in the muscle-fibre.

Dr. G. Thin* appears to be the first observer who described an intracellular network in the fibre of striped muscle. He examined frog's muscle treated with gold chloride in the following manner. After staining with gold chloride the muscle was exposed to light in acidulated water and then kept in strong acetic acid, at a temperature of 38°C., from 6—24 hours. By this method he demonstrated a network of fine fibres, concerning which he says, "this network was composed of exceedingly fine fibres, and its meshes accurately corresponded to Cohnheim's areas" (p. 252). He states that he demonstrated in the muscle-fibre, by the process of isolation, (1) the existence of flat cells (the muscle-corpuscles), (2) a network connected with central cellular protoplasm, and (3) parallel rows of spindle elements. Further on he states that he was "compelled to associate the transverse markings with the existence of this network, without attempting to explain the connection between them more definitely" (p. 258).

Gerlach† has written two papers on this subject. In the first paper he states that the contractile contents of the sarcolemma are traversed by a retiform substance continuous with, and identical with the axis cylinder of the nerve. He thus regards the network as of a nervous nature.

In the second paper he states that an intravaginal nerve network is present within the sarcolemma; and that in gold specimens striæ are seen which behave in a manner similar to the intravaginal network, and can be traced into continuity with it. He divides muscle into an anisotropous contractile matrix, and an isotropous nerve network. His results were obtained by the following method: The gold preparations were left several days in a mixture of 1—2 parts hydrochloric acid, 20 parts glycerine, and 20 parts water. This method brought out the longitudinal striæ. On making gold preparations as above, and subsequently treating with 1 per cent. potassium cyanide, he states that the sarcolemma gives way and the contents escape, partly as fine particles and partly in

---

* "On the Structure of Muscular Fibre," "Quart. Journ. Micr. Sci.," vol. xvi. (N. S.), 1876, pp. 251—259.

† "Das Verhältniss der Nerven zu den Muskeln der Wirbelthiere," Leipzig (Vogel), 1874. "Ueber das Verhältniss der nervosen und contractilen Substanz des quergestreiften Muskels," "Arch. f. Mik. Anat.," Bd. xiii., 1887, p. 399.

larger pieces; in such pieces the transparent substance bounding Cohnheim's areas was stained red, and was thickened at the nodal points. The muscle-corpuscles always lay in the stained substance and not in Cohnheim's areas; they consisted of a central oval nucleus and a stained peripheral substance continuous with the stained network of longitudinal striae. He states that the longitudinal striae are very variable in thickness and always zigzag: never straight. He regards them as thickenings of fine sheaths of nervous matter enclosing the fibrillae of the muscle; these sheaths corresponding to the boundaries of Cohnheim's areas.

He therefore concludes that the intravaginal nerve plexus and the longitudinal striae are continuous, and together make up the isotropous part of the muscle-fibre, and are to be considered as nervous. To them belong also the muscle-corpuscles and the nuclei of the intravaginal nerves. He regards the anisotropous matrix as the contractile part. Gerlach thus appears to view the isotropous part of the muscle, stained by the gold, as a honeycomb, and not a true network of fibrils. He has apparently failed to observe the transverse networks, and does not attempt to explain the relation between the network and the transverse striation.

The existence of an intravaginal nerve plexus in the muscle-fibre, and also the continuity of the nerve end plate with the isotropous part of the muscle, have been denied by Ewald* and Fischer.†

Engelmann‡ regarded the isotropous part of the fibre as a structure " das in physiologischer Hinsicht von einem Nerven nicht wesentlich abweichen würde," and suspected a connection with the axis cylinder.

In a later paper Engelmann§ states that the contractility of the muscle-fibre is always connected with fibrillar elements in the fibre; and he compares these with fibrillar elements in the protoplasm of some of the Rhizopoda, Infusoria, etc.

Retzius ∥ describes very carefully a network in the muscle-fibre of

---

* " Arch. für Mikr. Anat.," Bd. xiii., pp. 305—390.
† " Pflüger's Archiv," Bd. xii., pp. 529—548.
‡ " Pflüger's Archiv," Bd. xi., p. 462.
§ " Pflüger's Archiv," Bd. xxv., 1881, pp. 538—565.
∥ " Zur Kenntniss der quergestreiften Muskelfaser," " Biologische Untersuchungen," 1881, pp. 1—26, Pls. i., ii.

Dytiscus and other forms. This network consisted of (I.) transverse networks placed at regular intervals, and corresponding in position to Krause's membranes; (II.) longitudinal bars parallel to each other, and apparently running the whole length of the muscle-fibre, and connected with the transverse networks. His results were obtained partly by transverse and longitudinal sections, and partly by teased preparations. He also employed the following method of gold staining: the specimens were placed for twenty-five minutes in $\frac{1}{5}$ to $\frac{1}{2}$ per cent. gold chloride, either with or without previous immersion in 1 per cent. formic acid; then in 1 per cent. formic acid for ten to twenty hours, and exposed to the light. He gives the following description of the muscle-fibre of Dytiscus: In the axis of the fibre there are one or more rows of muscle-corpuscles, the protoplasm of which is produced into several (2—5) processes, from which finer processes arise forming the *transverse networks*. Each muscle-corpuscle is in connection with five or six successive transverse networks. The *longitudinal bars* of the network he describes rather doubtfully as consisting of rows of dots (p. 8), but he describes and figures them projecting freely in some preparations. The *matrix* is structureless, and is only slightly stained by the gold. The *sarcolemma* is apparently closely attached to, but probably independent of, the network. The *nerve-endings* appear to be in close connection with the transverse networks. The *function* of the network he was unable to determine, but states that it is probably not merely a supporting framework, but actively concerned in contraction. He does not, however, regard it as the true contractile part, as, according to him, it does not undergo important changes in form during contraction and extension. He thinks that the network is probably concerned in conveying the stimulus from the nerve to the muscle-fibre. In support of this he mentions that the fibre, as a rule, contracts simultaneously throughout its whole thickness; also that the nerve-fibres are apparently connected with the transverse networks.

Retzius also examined muscle from Musca, Oestrus, Notonecta, Locusta, Astacus, Rana, and Triton. In Locusta the transverse networks had more polygonal meshes than in Dytiscus. In Astacus, Rana, and Triton the longitudinal bars of the network were thicker than the transverse.

In some cases he states that the longitudinal bars of the network were not straight but zigzag. From the descriptions and figures it is very probable that this appearance was due to pressure, and is not normal.

Bremer* also describes a well-defined network in the striped muscle-fibre, evidently identical with that described by Retzius. He states that the longitudinal lines are true fibrils, and not part of cylindrical sheaths, as Gerlach maintained. He further traces the axis cylinder of the nerve into direct continuity with the muscle-corpuscles.

He considers the longitudinal striæ of Gerlach to be identical with the longitudinal bars of the network, and explains the irregular dotted appearance, or "Sprenkelung," of Gerlach's longitudinal striæ as being due to imperfect staining.

Bremer's results, though published subsequently to those of Retzius, were obtained quite independently.

B. Melland† has recently investigated the structure of the striped muscle-fibre, and has arrived independently at results agreeing very closely with those of Retzius and Bremer. This close correspondence between the accounts of these three observers affords satisfactory evidence of the correctness of their observations.

The points of difference between the networks described by Melland and Retzius are slight, the chief one being that Melland figures the transverse networks in Dytiscus with more polygonal meshes, and furnished with nodal thickenings at the points of junction with the longitudinal bars of the network. Retzius figures these in Locusta, but in Dytiscus he describes the transverse networks as generally composed almost entirely of radial fibres with very few transverse connections; and in place of nodal dots he describes several thickenings or nodes placed irregularly, and much fewer in number than the nodal dots described by Melland. Melland does not trace any connection between the network and the muscle-corpuscles, nor with the nerve-endings. He shows how the optical appearances of striped muscle are caused by the network. He considers the network to be intimately connected with the sarcolemma,

---

* " Arch. für Mikr. Anat.," Bd. xxii., 1883, pp. 318—356.

† "A Simplified View of the Structure of the Striped Muscle-fibre," " Quart. Journ. Micr. Sci.," July, 1885.

and to be homologous with the intracellular networks which have been described in other cells. It is evident from Fig. 6 of his paper that we have to do with a true network and not a honeycomb, a fact which is not so apparent from the figures of Retzius and Bremer. I have reproduced the figure from Melland's paper (Fig. 18).

Melland's results were obtained partly in conjunction with myself, and the object of the present paper is a continuation of this investigation. I have endeavoured to trace the distribution of this intracellular network of the striped muscle-fibre in the animal kingdom, and also, as far as possible, to determine its function.

The striation of muscle must not be confounded with a transversely striated appearance caused by a corrugated outline of the fibre, possibly due to a state of over-contraction. Such a false striation is met with occasionally in some fibres in Echinus, Hirudo, etc., and is the cause of the muscles of these animals having been described as striped. I shall, therefore, only describe muscle as being striped when the striation is due to the presence of the intracellular network described by Retzius, Bremer, and Melland.

I have examined muscle taken from representatives of the chief groups of the animal kingdom with the special object of investigating the presence of an intracellular network in the muscle-cells, either such as that of the striped muscle-fibre, or, when this does not exist, an intracellular network of any kind.

Amœba and Hydra have been included in this investigation; for it is an important point to determine the existence of an intracellular network in such a primitive and eminently contractile cell as Amœba: it is also important to investigate the structure of the muscular processes of the ectoderm cells of Hydra, as they are supposed to represent the first beginning of a muscle-cell.

In all cases the outlines and main details of the figures were drawn with the camera; in most cases under the $\frac{1}{10}$th immersion objective of Beck with No. 2 eyepiece, giving a magnifying power of 1,100 diameters.

Methods of Preparation.—The chief method of preparation used was the method of gold staining employed by Melland. The gold stains and renders evident the intracellular network of most cells, especially the network of the striped muscle-cell; hence it is at once a test whether the striation of the fibre is due to the presence of the network, or whether it is merely the false striation mentioned above.

Various modifications of the gold method were employed, according to the delicacy of the tissue under investigation. The method employed by Melland consists in placing the muscle in 1 per cent. acetic acid for a few seconds; then in 1 per cent. gold chloride for thirty minutes; then in formic acid, 25 per cent., for twenty-four or forty-eight hours in the dark.

This answers well for vertebrate and insect muscles. But for the more delicate organisms, such as Hydra, Daphnia, etc., and the heart-muscle of invertebrates, I found a one hour's immersion in formic acid, exposed to strong sunlight, to be the best treatment; or, in some cases, a warm chamber (40°C.) was used. A longer immersion than one or two hours in the formic acid in these cases leads to disintegration of the tissues. In addition to the gold preparations, osmic acid preparations were made in most cases, and compared with those made by the gold method. Osmic acid is well known for its property of fixing the histological elements in their natural state.

The examination of fresh tissues was in many cases of very little use; for the cells of the striped muscle of many of the animals investigated are so small that under a high power they barely appear striped, and no network can be seen at all. In these cases it is only by softening the fibre and so swelling it out, and at the same time staining the network, that the latter can be demonstrated; this is the special action of the method of gold staining.

It is necessary to mention that the results obtained by the gold method are somewhat uncertain. In some cases the network will come out distinctly, but in others, especially when the preparation has been left for a longer time than usual in the acetic acid, the network appears to consist of rows of granules instead of definite lines. This uncertainty was noticed also by Retzius, Gerlach, and Bremer, and is no doubt the cause of the different appearances described by these authors.

In order to avoid the monotony and interruption of repeatedly stating the treatment used for the muscle of each animal, the exact method used is given with the description of the figure of each animal in the plates at the end of the paper.

I shall now take the chief groups of the animal kingdom in their zoological order, describing the muscle found in each case.

STRUCTURE AND DISTRIBUTION OF MUSCLE IN THE ANIMAL KINGDOM.

PROTOZOA.

Amœba.—Klein* states, on Heitzmann's authority, that under suitable conditions the protoplasm of the white blood corpuscles can be seen to contain an intracellular network composed of fine fibrils. Dr. Klein has, however, recently informed me that he does not find an intracellular network in Amœba, nor in the majority of white blood corpuscles.

On examination of very large specimens of *Amœba princeps* in the fresh state, the constant flowing movement of the protoplasm renders it difficult to conceive of any permanent intracellular network. I have, moreover, made gold preparations of these Amœbæ in the following manner: The Amœba was placed in a drop of water with a little cotton wool underneath the cover glass to prevent the animal being washed away by the reagents. A few drops of 1 per cent. acetic acid were then run in under the cover glass for a few seconds. Gold chloride was then run in, and the animal left in this for fifteen minutes. Formic acid was then added, and the animal left exposed to light for about one hour; by this time the gold was reduced and the animal stained. The preparation was then mounted in dilute glycerine.

Amœbæ prepared in this way showed no trace of an intracellular network; the protoplasm simply presenting a mottled granular appearance.

Although there is no definite intracellular network, comparable to that of an ordinary epithelial or gland cell, known to exist in any of the Protozoa, yet a vacuolated condition of the protoplasm is well known to occur in many of them. This attains a high degree of development in many forms, *e.g.*, Noctiluca. These vacuoles are certainly not all food vacuoles, and may possibly indicate the starting point of the differentiation of an intracellular network, *i.e.*, a differen-

---

* "Atlas of Histology," p. 2, diag. 1.

tiation of the cell into firmer and less dense parts, the former of which takes on the form of a network or reticulum. For although it is not absolutely certain that the structures described as intracellular and intranuclear networks are in all cases denser than the rest of the protoplasm of the cell, they are, I believe, generally assumed by histologists to be so, and also to be protoplasmic in nature.

Vorticella.—The stalk of the Vorticella contains a spiral protoplasmic fibre, which is eminently contractile. This fibre, when treated with the gold staining, shows no trace of the presence of fibrils, having simply the appearance of undifferentiated protoplasm.

CŒLENTERATA.

Hydra.—The peculiar ectoderm cells of the Hydra are important to investigate, since they are generally held to represent the first commencement of a muscle. Here the one cell is differentiated into two parts to perform two functions, the one portion to act as a sensory cell, the other to act as a muscle.

Hamann[*] describes, in the epithelial muscle-cells of the hydroid polypes, a network in the body of the cell, but no fibrillation in the muscular process.

My own observations on the cells of the Hydra agree with those of Hamann. Gold preparations of these cells show a network in the body of the cell, but no continuation of it into the muscular process (Fig. 1).

Medusa.—Striated muscle has been described as occurring in the disc of Aurelia by Max Schultze, Brücke, and Virchow, and in Pelagia by Kölliker.[†]

In gold preparations of muscle from the disc of Aurelia I find distinct transverse striation, which, under the $\frac{1}{18}$ immersion objective, is found to be due to the presence of a network similar in all respects to the network described by Retzius and Melland in striped muscle (Fig. 2).

Actinia.—Muscle taken from the base of the Actinia and treated with gold was found to consist of elongated fusiform cells, non-

---

[*] "Organismus der Hydroid Polypen," p. 15.
[†] "Stricker's Handbook of Histology," vol. iii., p. 551.

striped, and showing no trace of any intracellular network, or of any fibrillation.

Hence the conclusions obtained are that in the muscular process of the Hydra cell there is no form of network or fibrillation, although a network is present in the body of the cell. In the more highly organised Medusa the typical network of striped muscle is found to be present, but in the equally highly organised Actinia no network is present, nor is there any fibrillation in the muscle-cells.

These results agree with those obtained by Hamann* by the method of isolating the cells by maceration in various reagents. He states that the muscles of Hydroid polypes are always smooth, and quite distinct from the striated muscles of the Medusæ. Hamann thinks that where transverse striation has been described in Hydroid polypes, it is probably due to the action of reagents.

The Hertwigs,† in their observations on the Actiniæ, describe no fibrillation in the muscle-fibre. They investigated the tissues of Sagartia and Anthea.

ECHINODERMATA.

The muscle of the Echinoderms has been described as striped by several observers. Firstly, by Schwalbe‡ in the muscle-cells between the ambulacral plates of *Ophiothrix*, and more recently by Geddes and Beddard.§ From the figure given by the latter observers it is evident that the striation they describe is the false striation mentioned above as being due to annular constrictions. Schwalbe, however, describes double oblique striation.

I have made gold preparations of muscle taken from the "lantern of Aristotle" of *Echinus*, and find no trace of the network of striped muscle or of any fibrillation. The cells are remarkable for the clearness and transparency of their protoplasm.

These results again agree closely with those obtained by Hamann.|| He describes the muscle of the Asterids as smooth, and very seldom

---

* Loc. cit., p. 20.
† "Die Actinien.".
‡ "Archiv. für Mik. Anat.," 1869, p. 205.
§ "Proc. Royal Soc. Edinburgh," 1873.
|| "Asteriden," p. 94, plate iii.; "Holothurien," p. 38, pl. ii.

showing fine longitudinal fibrillation. In the Holothurians he describes the muscle as non-striped, but states that longitudinal fibrillation is to be seen in it.

Vermes.

Hirudo.—The muscle-fibres of the Leech are peculiar; they consist of an outer clear portion and a central granular part. In gold preparations the outer part stains the more deeply of the two portions of the cell, and appears quite homogeneous, showing no trace of a network. In osmic acid preparations the outer layer appears very faintly fibrillated, but I could not identify any distinct fibrils differentiated from the rest of the cell even under the $\frac{1}{10}$ immersion objective.

Transverse sections of the muscle of the Leech show a radiating appearance of dark and light bands in the outer portion of the cell. This is, I believe, caused by the method of preparation, for in some sections the outer portion of the cell is broken up into pieces arranged in a radiating manner and corresponding to the light portions between the radiating dark lines in the better preserved specimens. I find nothing corresponding to this appearance in muscle prepared by the gold or osmic acid methods, which are the methods generally recognised as maintaining the true histological characters of cells intact.

Wagener,[*] from transverse sections of dried specimens of Leech, states that the muscle-cells consist of a central medullary substance round the nucleus, and a cortical substance splitting into fibrils. This is also described by Schwalbe in the fibres of *Aulastomum*.

Lumbricus terrestris.—Gold preparations of the muscle of the Earthworm show large elongated cells which on close examination show longitudinal lines; these under the $\frac{1}{10}$ immersion objective present a dotted appearance (Fig. 3). At first sight it might appear that we have here the network of striped muscle; but this is not the case. In the first place, there is no appearance of transverse striation at all; in the second place, the dots are not arranged transversely but are quite irregular; lastly, so far as I could observe,

---

[*] "Archiv. f. Mik. Anat.," 1869.

the dotted lines are superficial, and do not extend into the body of the cell.

One of the so-called "hearts" of the worm was treated in the same way; the muscle-cells were found to resemble almost exactly the muscle-cells described above.

In the Polyzoa and Rotifera striped muscle is well known to occur. I have not, however, been able to determine with certainty whether the striation is due to the presence of a network or not. Nitsche* says that the striation in the retractor muscle of the Polyzoa is not due to any wrinkling of the sarcolemma. The retractor muscle appears to be the only muscle that is striped, from Nitsche's observations.

Striped muscle has been recently described by Haswell† in the gizzard of Syllis, one of the Polychæte worms.

MOLLUSCA.

According to Schwalbe,‡ double oblique striated muscle is present in Solen, Ostrea, and Helix.

Anodon.—Preparations of the adductor muscle of the Anodon treated with gold show that the muscle consists of small elongated cells of the unstriped type, showing no fibrillation or transverse striation. The muscle treated with osmic acid shows faint fibrillation, but no distinct fibrils.

Patella.—The Limpet was chosen for an investigation of the structure of the heart muscle. In gold preparations of the ventricle I find the network of striped muscle present (Fig. 5).

In the heart of the Anodon I could not determine with certainty whether the network was present or not, although faint indications of it were obtained.

Ostrea.—The adductor muscle of the Oyster consists of two portions: a white opaque portion, and a more gelatinous portion. Gold preparations were made of each of these. The cells of the "white muscle" are large, with clear outlines, and remarkable for the clearness and transparency of the protoplasm. The cells of the "gelatinous muscle" are smaller and less transparent. Neither of these showed any network or fibrillation.

---

* "Kenntniss der Bryozoen," Heft 2, p. 55.
† "Quart. Journ. Micr. Sci.," 1886.
‡ "Arch. f. Mik. Anat.," 1869.

Helix pomatia.—Gold preparations of the muscle of the foot show that it consists of very small cells of the unstriped type densely massed together.

The muscle of the odontophore, however, shows transverse striation, which under the high power is seen to be caused by the presence of the typical network of striped muscle.

Pecten.—The Pecten differs from most of its class by performing rapid movements of its adductor muscle, whereby it propels itself through the water. Gold preparations of the adductor muscle made by my friend, Mr. J. T. Cunningham, show the network of striped muscle very plainly (Fig. 4). I have not observed the double oblique striation described by Schwalbe in the muscle of Molluscs and Echinoderms. As this is not seen in gold and osmic acid preparations, I think it must be an optical effect. Schwalbe, indeed, admits that in Ophiothrix the transverse striation is due to folds in the sarcolemma (loc. cit., p. 211).

ARTHROPODA.

Representatives of the Crustacea and Insecta, viz., the Lobster, Dysticus, and the Bee, were investigated by Mr. Melland,* the network, of striped muscle, being found in each case.

Astacus, Heart-muscle.—Gold preparations of the heart of the Crayfish show the network to be present in this muscle as in the body-muscles; the network is, however, much finer and more difficult to demonstrate (Fig. 6).

The muscle-fibres of the heart are intimately blended with what appear to be large masses of granular protoplasm enclosing nerve-cells; these may possibly be of the nature of nerve-endings.

Daphnia.—As a representative of the minuter forms of Crustacea, I examined the Daphnia. The muscle-fibres of this animal, when examined in the fresh state, only show transverse striation faintly. After many attempts I succeeded in obtaining a satisfactory gold preparation, where the muscle-fibres were much softened and pressed out to many times their normal diameter. These fibres show the network very plainly (Fig. 7).

In this case the animal was placed whole in 1 per cent. acetic acid

---

* Loc. cit.

for ten minutes, and left in the formic acid in a warm chamber at 40°C. for two hours.

Insect Larva.—To determine if striped muscle is present in the larval insect as well as in the imago, I made preparations from the larva of the Ermine Moth (*Spilosoma lubricepeda*). Muscle was taken both from the jaws and from the legs. In both cases the muscle was found to be striped, the network in the muscle of the jaw being especially well developed.

Arachnida.—Muscle taken from the leg of the Spider and treated with gold showed the network of striped muscle.

VERTEBRATA.

The vertebrate animals examined by Mr. Melland were the Frog and the Rat. These will serve as examples of the Amphibia and Mammalia. I have examined the muscle of animals taken from the other chief groups, viz., the *Cyclostomata, Elasmobranchii, Teleostei, Reptilia,* and *Aves,* taking as representatives of these groups respectively: Myxine, Scyllium, Gastrosteus, Testudo, and Turdus.

Muscle taken from each of these animals was treated with the usual gold method, and in each case a network was found identical with that described by Melland. On comparing these networks with one another and with those described above in the striped muscle of the several invertebrate animals, they are found to agree in all respects.

With regard to *Amphioxus,* I have not had the opportunity of examining fresh specimens. The muscle has, however, been described as striped,* and, from analogy, I see no reason to think that the striation is due to any other cause than a network. In the *Ascidian,* I have examined the muscular bands of Salpa, and find striped muscle present.

CARDIAC MUSCLE OF VERTEBRATES.

The heart-muscle has long been described as faintly striped transversely, but whether this striation is due to the same cause as that to which ordinary striped muscle owes its striation, has not been determined with certainty.

---

* Grenacher, "Zeit, für wiss, Zool.," p. 577.

In order to investigate this point, I made gold preparations of muscle taken from the Rat's heart. The cells are seen to contain a network similar to that of ordinary striped muscle. The network is more delicate and with much smaller meshes than the network in the body-muscle of the same animal, and is therefore more difficult to demonstrate by the gold method (Fig. 9). I have also prepared muscle from the heart of the Frog (Fig. 10) and Bird (Fig. 11); the network in the latter animal is much plainer than in the others.

The striation of cardiac muscle therefore appears to be due to an intracellular network similar to that of ordinary striped muscle.

Unstriped Muscle of Vertebrates.

Klein* describes in the unstriped muscle of vertebrates a bundle of longitudinal fibrils which are in connection with the intranuclear network. His description of the structure of the unstriped muscle-cell is as follows: "Thus we may regard the unstriped muscle-fibre as composed of a sheath with annular thickenings and a bundle of delicate fibrils which at one more or less central point forms a delicate network. This, surrounded by a special membrane—except where the network is in connection with the bundle of fibrils—represents the nucleus." Klein regards the bundle of fibrils as the contractile part of the cell, and thinks that by their shortening the muscle-fibre is caused to contract, elongation being produced by the elastic rebound of the sheath. Flemming† has observed these fibrils in the living muscle.

Dr. Klein informs me that he regards the bundle of longitudinal fibrils as representing an intracellular network homologous with other intracellular networks. Dr. Klein has been kind enough to show me his preparations of unstriped muscle prepared from the mesentery of the newt by twenty-four hours immersion in 5 per cent. ammonium bichromate, and afterwards stained with logwood. These preparations show clearly the longitudinal fibrils and their connection with the intranuclear network.

I have made preparations, by the gold method, of muscle from the

---

* Klein, " Atlas of Histology," p. 74, pl. xv.; also "Quart. Journ. Micr. Sci.," 1878.

† " Beobachtungen über die Beschaffenheit des Zell Kerns," "Arch. für Mikr. Anat.," 1876, Bd. xiii., pp. 714, 715.

mesentery of the newt and from the bladder of the Salamander. In both of these the fibrils in the muscle-cells are very evident, but the intranuclear networks do not show at all distinctly, which is a most unusual result in this mode of preparation. However, in preparations from the mesentery of the newt, made by Klein's method, the intranuclear networks come out very distinctly in many fibres; and in one case I could trace the connection of the intranuclear network with the fibrils of the cell. The longitudinal fibrils do not show so well in these preparations as in those made by the gold method (Figs. 12, 13). It thus appears that the vertebrate unstriped muscle differs from all the invertebrate unstriped muscle that I have investigated, in that the cells contain an intracellular network in the form of longitudinal fibrils. This may perhaps represent a form of network intermediate between the typical irregular network of other cells and the highly modified network of the striped muscle-cell.

From these investigations it appears that the peculiar intracellular network of striped muscle is developed in all muscles which have to perform rapid or regular movements.

A brief review of the chief animals mentioned in the preceding pages will make this clear. Commencing with the Actinia and the Medusa, these are both highly organised Cœlenterates, but the Actinia is a sluggish animal, which exhibits slow and irregular movements, while the Medusa propels itself through the water by rapid and regular contractions of its disc. Now, in the Actinia we find no striped muscle, but in the Medusa the network is present. In the worms, such as the Leech and Earthworm, striped muscle is absent; these animals only performing comparatively sluggish movements. In the Polyzoa the retractor muscles of the stomach, and in the Rotifers the retractors of the trochal disc, perform rapid movements, and have been described as striated transversely; this is probably due to the network, although, as stated above, I have so far been unable to determine this myself. In the Mollusca the movements are as a rule sluggish, and unstriped muscle is the prevailing type in this group. But in the odontophore muscles of the Snails the movements are more rapid, and in these we find the network developed. Also in the hearts of these animals, which perform rapid and regular contractions, I find the network present, at any rate in the case of Patella.

In the Pecten we have a Mollusc which differs from the majority of its class by performing rapid movements by the contraction of its adductor muscle, and here we find the network present. This is a most important fact in favour of the view that the peculiar network of striped muscle is developed when rapid movements are to be performed; for here we have the Mussel and the Pecten, both belonging to the same division of the Mollusca, and both having adductor muscles moving the valves of the shell. In the Mussel the adductor muscles only act at irregular intervals, and comparatively slowly; but in the Pecten they perform rapid and frequent contractions when the animal swims. In the Mussel we find unstriped muscle, but in the Pecten the network of striped muscle is present.

In the majority of Arthropods and Vertebrates the movements are chiefly rapid and of frequent occurrence, and in these groups there is a wide distribution of striped muscle.

It is quite possible that in some animals of sluggish habits, such as some adult insects, the presence of striped muscle may be due to inheritance.

We should expect on this view to find striped muscle present in all well-developed hearts, since they execute rapid and regular contractions. However, in the so-called "hearts" of the Earthworm the muscle is unstriped. This can, I think, be explained as follows. These so-called "hearts" represent the earliest and most primitive form of heart in the animal kingdom, being simply local hypertrophies of the blood vessels which perform rhythmic contraction. Now, the muscle of the blood-vessels is unstriped, therefore we should scarcely expect to find striped muscle in what are simply local hypertrophies of those vessels. Moreover, the contraction of these "hearts" is slow and peristaltic in nature. It is only when we come to the more highly developed hearts, such as those of the Patella, Snail, &c., which have to perform much more rapid and regular contractions than the "hearts" of the worm, that we find striped muscle developed.

I may here state that I have not yet been able to determine the nature of the connection between the network of striped muscle and the nerve end-plate, which must exist if the combined results of Retzius and Bremer are correct. This I hope to do in a subsequent paper. I have, however, recently observed the connection between the network and the muscle-corpuscles described by Retzius.

Theory of Muscular Contraction.

The general conclusions arrived at in the preceding part of the paper are as follows:—

(1) An intracellular network of a definite character is present in the fibre of striped muscle throughout the animal kingdom.

(2) This network is developed where rapid and frequent movements have to be performed.

(3) The striped muscle-fibre consists of sarcolemma, network, and sarcous substance; and, so far as at present determined, there is no other structure present in the fibre (except the muscle-corpuscles and nerve-endings).

The question now before us is to determine if possible the nature and function of the network, and what relation it bears to the contractility of the muscle-fibre.

Changes in the Network during Contraction.—In order to investigate this point I teased out some perfectly fresh muscle from the leg of a Dytiscus and placed it on an inverted cover-glass over a gas-chamber. Alcohol vapour was then blown over the preparation, when most of the fibres contracted owing to the chemical stimulus. The vapour was passed over the muscle for about a quarter or half a minute. The fibres were then fixed in their contracted state by plunging them into 5 per cent. acetic acid for half a minute, and then treated with gold and formic acid in the usual way. Many fibres were thus obtained completely contracted, and also many fixed waves of contraction.

I also made preparations of relaxed muscle from a Dytiscus killed with chloroform. However, as the fibres vary so much in appearance, according as they are more or less pressed out in the gold preparation, comparisons of the muscle stimulated with alcohol vapour, with that reduced by chloroform, though they may give the general effect of the difference, are not absolutely trustworthy. The only way of really proving this point is to examine a fibre, one portion of which is in the relaxed condition and the other contracted, or in other words, a fixed wave of contraction.

On careful examination of the network in one of these fixed waves of contraction with the $\frac{1}{10}$ immersion objective, the longitudinal fibrils of the network were *always straight in all parts of the fibre*, and appeared *slightly thicker in the contracted part* of the fibre, although it was difficult to judge accurately of the difference in thickness. *The nodal dots, however, were the same size in both the contracted and relaxed portions of the fibre.* The dots appeared in many cases even smaller in the contracted than in the relaxed muscle. This is, I believe, due to their being more separated from each other laterally, whereby the refractive effects which somewhat obscure the real size of the dots in the relaxed muscle are diminished (Fig. 14).

It therefore appears from gold preparations that during contraction the nodal dots do not alter in size, but that the longitudinal bars of the network increase in thickness. The apparent enlargement of the nodal dots when the fibre is seen in the fresh state is due to optical effect. Moreover, if the nodal dots do not alter in size, it follows necessarily that the longitudinal bars must increase in thickness; for since they keep straight during contraction, if they do not increase in thickness there must be a diminution in the volume of the fibre, which is known not to occur.

These results differ from the account given by Schäfer of the changes during contraction. He states[*] from observations on the living fibre that during contraction his "muscle-rods" (which correspond to the longitudinal bars of the network) become compressed in the centre, and their substance tends to accumulate towards the ends, *i.e.*, that the knobbed ends of the muscle-rods, which correspond to the nodal points of the network, increase in size at the expense of the shafts connecting them. On examination of the living fibre this certainly appears to be the case, but the optical effects of reflection and refraction are so great as to obscure the real change that takes place.

Nature and Function of the Network.

In discussing the theory of contraction, I shall assume that the intracellular network of striped muscle and the longitudinal fibrils of the vertebrate unstriped muscle, are of the same nature as other intracellular networks; and, in accordance with the views of modern

---

[*] "On the Leg-muscles of the Water-beetle," "Phil. Trans.," 1873.

histologists, that they are protoplasmic in nature, and denser than the rest of the cell.

We have first to consider the nature of intracellular networks in general, and whether their function is an active or a passive one. In the case of intranuclear networks, the changes which the network undergoes in karyokinetic division of the nucleus point to their being of an active nature. The extranuclear network (intracellular) is apparently of the same nature as the intranuclear, since the two have been shown to be continuous in many cells; and also they have the same behaviour towards stains and reagents. Moreover, if intracellular networks are developed by a process of vacuolation of the protoplasm of the cell, or a division into denser and less dense parts, as described previously when treating of the Protozoa, it is obvious that in these cases the network must be the active and the contractile part of the cell.

The continuity and identity of nuclear and extranuclear networks is strongly supported by Sedgwick's remarkable observations on the early stages of Peripatus.* He not only demonstrates the continuity of the extranuclear and intranuclear networks, but he also shows that during segmentation of the ovum the cells do not become completely separated, but remain connected by their protoplasmic networks, i.e., that the intracellular networks of all the cells are continuous.

He also states that the so-called nuclear membrane is reticular in nature, and not a true membrane, being, in fact, part of the general reticulum of the cell. In the cells described by Sedgwick there is no doubt that the reticulum is the active portion of the cell, for the rest of the cell consists simply of vacuoles.

Flemming states,† as Sedgwick also noticed, that the first change observable in a cell whose nucleus is about to divide is in the extranuclear protoplasm. Strasburger‡ further states that the fibrils which form the nuclear spindle originate in the surrounding "cytoplasm" at the time of division. This appears to be direct evidence of an active function in the intracellular network.

---

* "Quart. Journ. Micr. Sci.," vol. xxvi., 1886, pp. 175—212.
† "Zellsubstanz, Kern u. Zelltheilung," Leipzig, 1882.
‡ "Arch. f. Mikr. Anat.," Bd. xxiii., "Die Controversen der indirecten Kerntheilung."

These considerations show that the function of intracellular networks is very probably of an active nature.

We have now to consider the networks of striped and unstriped muscle. Both these forms of network are non-essential to contraction, for we have seen that many muscle-fibres of invertebrates are devoid of a network of any kind; but that they modify the nature of the contraction is very probable.

We have seen that the network of striped muscle is developed when rapid movements are to be performed; this shows that the function of contraction is intimately associated with the presence of the network.

The chief points of difference in the contraction of striped and unstriped muscle respectively are the great length of the latent period and the long duration of the contraction in the unstriped muscle. The velocity of the contraction-wave in striped muscle is in the Frog, 3—4 metres per second, while in the unstriped muscle (ureter) the velocity is only 20—30 mm. per second.* This seems to indicate that the peculiarly arranged network of striped muscle may be associated with the rapidity of its contraction.

In nearly all the specimens I have examined, *both the transverse and longitudinal bars of the network remain perfectly straight in all conditions of contraction and relaxation of the muscle. Hence the network, or part of the network, must either contract to the full extent that the muscle-fibre does, or else be elastic, and so follow the movements of contraction of the fibre.*

Retzius† figures a specimen in which the longitudinal bars are zigzag. However, from his description, and from comparison with my own preparations, I believe this to be due to disturbance during the preparation, and not to be a normal condition.

We have now to consider whether the network is actively contractile or merely a passively elastic structure; or whether one part of it is contractile and the other passive. That both network and sarcous substance are contractile is improbable; for if the function of the network and the sarcous substance is identical, there is no apparent reason for the presence of the network. Differentiation in structure always implies differentiation in function.

* "Text-book of Physiology," Dr. Michael Foster, 4th ed., p. 101.
† Loc. cit., pl. i., Fig. 19.

Action of the Longitudinal Bars of the Network.

We have seen that the longitudinal bars of the network diminish in length and apparently increase in thickness during contraction, and *that they always remain straight in all conditions of contraction and relaxation of the fibre*. The question now before us is to determine if they are *actively contractile* or *passively elastic*. The following considerations are opposed to the latter view :—

(*a*) If the longitudinal bars of the network are passively elastic, they must be on the stretch in the relaxed condition of the fibre, and resemble stretched elastic threads running the whole length of the muscle-fibre. Now, when a muscle is cut out of the body, and thereby removed from its attachments, it does not contract to any considerable extent; therefore, supposing the longitudinal bars to be elastic, something must keep them on the stretch.

(I.) This cannot be the sarcous substance, for as it is semi-fluid in nature it can hardly keep elastic threads on the stretch.

(II.) It cannot be a nervous impulse, continually acting on the longitudinal bars, for if it were so a muscle would contract on section of its nerve.

(III.) The only force which can keep the bars on the stretch must be that of the transverse networks. On this supposition the uncontracted muscle is not in a state of rest, for there is a continual force exerted against the transverse networks by the tendency of the longitudinal bars to shorten. It is very difficult to conceive that the muscle, in its uncontracted condition, should be in a state of extreme tension, and not of comparative rest.

(*b*) In the unstriped muscle-fibre there are no transverse networks present, and hence no force to keep the longitudinal fibrils on the stretch, except the sarcolemma, which would be scarcely adequate to do so.

It therefore appears improbable that the longitudinal bars of the network are passively elastic, and if this is the case the only conclusion remaining is that they are actively contractile, and hence, presumably, the cause of contraction of the fibre. This view is also supported by the following considerations :—

In the muscle-cell the part which performs the contraction is evidently the most fundamental part of the cell, and this we should expect to be differentiated first. In the embryonic development of striped muscle it is found that the longitudinal striation appears first, *i.e.*, that the longitudinal bars of the network are differentiated before the transverse. This is also the case in regenerating muscle. Again, in tracing the phylogeny of muscle, we found that the first indication of an intracellular network was in the vertebrate unstriped muscle in the form of longitudinal bars only. Hence both the phylogeny and the ontogeny of the network favours the view that the longitudinal bars are the contractile part of the cell.

Action of the Transverse Networks.

Similarly to the longitudinal bars the transverse networks *always remain straight in all conditions of contraction and relaxation of the fibre.* Hence they become necessarily extended when the muscle-fibre contracts, and return to their original form on relaxation of the fibre. The question now remains as to whether the return of the transverse networks to their original position is due to *active contractility or to elastic rebound*. The following arguments, for the first of which I am indebted to Mr. Melland, are in favour of the latter view.

(*a*) An elastic thread, if stretched and then allowed to rebound, will always return to its original length, *i.e.*, will always shorten to the same extent. The transverse networks behave in this way; they always shorten to the same extent, viz., to the normal diameter of the fibre. This speaks in favour of their being passively elastic, for if they were actively contractile there is no reason why the fibre should not be compressed to less than its normal diameter, elongation at the same time taking place; whereas the fibre always relaxes to the same extent.

(*b*) If the statements of Gerlach, Retzius, and Bremer are correct, both parts of the network are connected with the end-plate and with the axis cylinder of the nerve, the longitudinal bars being connected indirectly through the transverse networks, the latter being in direct connection with the nerve. It is therefore difficult to conceive that the transverse networks can contract actively *after* the longitudinal bars have begun to relax, for the nervous impulse will apparently reach the former first, and hence they must contract at the same time as or *before* the longitudinal bars; and yet if the relaxation of the fibre is held to be due to active contraction of the transverse networks, this is what must occur.

Conclusion.

The conclusion to which I am therefore led is that the contraction of the striped muscle-fibre is due to the active contraction of the longitudinal bars of the network, and that the transverse networks are probably passively elastic, and by their rebound cause relaxation of the muscle-fibre. That the transverse networks and the muscle corpuscles, with which they are said to be continuous, possibly furnish paths by which the nervous impulse is conveyed from the nerve-ending to the longitudinal bars. That the contraction of the unstriped muscle-fibre is due to the active contraction of its longitudinal fibrils when these are present, as in vertebrate muscle. In the case of unstriped muscle which possesses no fibrils, the contraction is due to the whole protoplasm of the cell, there being no special part differentiated to perform this function.

Should these conclusions prove to be correct, we may imagine the changes that occur in the striped muscle-fibre during contraction to be as follows :—

The nervous impulse reaching the end-plate of the nerve is conducted by the transverse networks to the longitudinal bars, and causes them to shorten ; it does not cause the transverse networks to contract, because they are passively elastic and non-contractile. The longitudinal bars shorten according to the strength of the nervous impulse, and remain so as long as it lasts. By fluid pressure the transverse networks are extended, and remain so as long as the longitudinal bars remain contracted ; when these cease to contract the elasticity of the transverse networks comes into play, and they shorten to their original dimensions, and by fluid pressure extend the longitudinal fibrils to their original length, the elastic sarcolemma aiding in the process.

The alternate action of the longitudinal and transverse networks no doubt causes the special features of the contraction of striped muscle, viz., the quick response to stimulus and the rapid contraction ; and we have seen that the network is developed wherever rapid movements have to be performed.

In connection with the foregoing considerations, the results of

Gerlach, Retzius, and Bremer, should they prove to be correct, are of importance. I think there is little doubt that the longitudinal striæ described by Gerlach are identical with the longitudinal bars of the network figured by Retzius, Bremer, Melland, and myself. Gerlach traced these striæ into connection with the nerve-endings. Retzius showed the connection between the muscle-corpuscle and the transverse striæ, and Bremer traced the axis cylinder of the nerve into direct continuity with the muscle-corpuscles. It therefore appears that the network is connected with the nerve, and that the longitudinal bars are connected with it indirectly through the transverse networks. The direct continuity of the network with the nerve does not necessarily imply that the network is itself nervous; in fact, it really supports the view that it is the part actively concerned in contraction; for we should expect, *a priori*, that if a differentiation occurred in muscle, it would be with the contractile part that the nerve would be in continuity.

On the other hand, with regard to the transverse networks, it is possible that they may be in part nervous in nature, and have for their function the more rapid conveyance of the stimulus through the muscle; and that the more rapid response to stimulus, the special characteristic of striped muscle, may be partly explained in this way.

There are two obvious objections to the theory of contraction we have arrived at, which I shall proceed to discuss:—

1. It necessitates a difference between the longitudinal and transverse bars of the same network. This is an objection, the real nature of which it is impossible to determine in the present state of our knowledge of the nature and import of intracellular networks in general. In unstriped muscle the longitudinal fibrils are alone present, and in the development of striped muscle the longitudinal elements of the network appear first. The transverse networks are described and figured by Retzius as direct processes of the muscle-corpuscles; the mode of their development is as yet unknown, but should they prove on further investigation to develop as processes of the corpuscles, it would follow that the two elements of the network are, in spite of their close connection in the adult, of entirely independent and different origin. And then a difference of function would become not only possible but highly probable.

Further, the action of different reagents in splitting the fibre in different directions (alcohol, etc., causing longitudinal, and acids transverse splitting) lends some support to the same view. Haswell* in his observations on the striped muscle of the gizzard of Syllis, states that after treatment with hæmatoxylin, and then glacial acetic acid, the transverse networks are stained, but not the longitudinal; he says this may point to some difference in the substance of which they are composed.

2. This theory attributes the function of contraction to the network, which forms much less of the bulk of the fibre than does the sarcous substance, the latter being far greater in amount than the network. In reference to this it should be borne in mind that contraction is not the only function performed by muscle. The muscles, as stated by Dr. Michael Foster,† are continually undergoing metabolism, giving rise to a certain amount of heat; the metabolism during rest being slow, but suddenly increasing during contraction. The energy involved in the work done in a muscular contraction is only about one-tenth the total energy expended, the rest going out as heat. Hence the muscles must be regarded as the chief sources of heat of the body, and are, "*par excellence*, the thermogenic tissues."

It thus appears that the thermogenic function of muscle absorbs a far greater amount of its energy than does the contractile function, and if we attribute the thermogenic function to the sarcous substance, and the contractility to the network, the above objection appears to receive a satisfactory answer.

The following quotation from Dr. Michael Foster‡ is curiously in accordance with the view of the structure and function of muscle maintained above, and may fitly conclude this paper.

"It is quite open for us to imagine that in muscle, for instance, there is a framework of more stable material, giving to the muscular fibre its histological features, and undergoing a comparatively slight and slow metabolism, while the energy given out by muscle is supplied at the expense of more fluctuating molecules, which fill up, so

---

* "Quart. Journ. Micr Sci.," 1886.
† "Text-book of Physiology," 4th ed., p. 461.
‡ Loc. cit., p. 475.

to speak, the interstices of the more durable framework, and the metabolism of which alone is large and rapid."

Summary.

1. In all muscles which have to perform rapid and frequent movements, a certain portion of the muscle is differentiated to perform the function of contraction, and this portion takes on the form of a very regular and highly modified intracellular network.

2. This network, by its regular arrangement, gives rise to certain optical effects, which cause the peculiar appearances of striped muscle.

3. The contraction of the striped muscle-fibre is probably caused by the active contraction of the longitudinal fibrils of the intracellular network; the transverse networks appear to be passively elastic, and by their elastic rebound cause the muscle to rapidly resume its relaxed condition when the longitudinal fibrils have ceased to contract; they are possibly also paths for the nervous impulse.

4. In some cases where muscle has been hitherto described as striped, but gives no appearance of the network on treatment with the gold and other methods, the apparent striation is due to optical effects caused by a corrugated outline in the fibre.

5. In muscles which do not perform rapid movements, but whose contraction is comparatively slow and peristaltic in nature, this peculiar network is not developed. In most if not all of the invertebrate unstriped muscle there does not appear to be an intracellular network present in any form, but in the vertebrate unstriped muscle a network is present in the form of longitudinal fibrils only; this possibly represents a form of network intermediate between the typical irregular intracellular network of other cells and the highly modified network of striped muscle.

6. The cardiac muscle-cells contain a network similar to that of ordinary striped muscle.

The investigations connected with this paper were partly carried on in the laboratories of the Owens College and partly at the Scottish Marine Station at Granton. I must here express my thanks to my brother, Professor Milnes Marshall, for his kindness in revising the paper, for much advice in its production, and for obtaining the literature of the subject; all the controversial points were discussed with him, and the preparations submitted to his examination. My thanks are also due to Dr. Klein for kindly showing me his preparations and for examining several of my own. I must also thank Mr. J. T. Cunningham for the use of the Scottish Marine Station, and for obtaining several of the animals.

DESCRIPTION OF PLATE VI.

Illustrating Mr. C. F. Marshall's paper on "Observations on the Structure and Distribution of Striped and Unstriped Muscle in the Animal Kingdom, and a Theory of Muscular Contraction."

In all cases the gold chloride used was 1 per cent., the formic acid 25 per cent. The "usual gold method," when stated, means 1 per cent. acetic acid for a few seconds, gold chloride thirty minutes, formic acid twenty-four hours in the dark.

Fig. 1. Ectoderm cell of Hydra: gold preparation. (1 per cent. acetic acid for a few seconds, gold chloride thirty minutes, formic acid one hour, exposed to sun.) a. Intracellular network. b. Muscular process. c. Intranuclear network. $\frac{1}{10}$ immersion obj.

Fig. 2. Portion of muscular fibre of Medusa. Usual gold method. $\frac{1}{10}$ imm. obj.

Fig. 3. Portion of muscle-cell of Earthworm, showing longitudinal rows of dots. Usual gold method. $\frac{1}{10}$ imm. obj.

Fig. 4. Muscle-fibre from adductor of Pecten, showing network. Usual gold method. $\frac{1}{10}$ imm. obj.

Fig. 5. Muscle-fibre from heart of Patella. (Acetic acid, 5 per cent., two minutes, gold thirty minutes, formic acid two hours, at 40°C.) Zeiss, J obj.

Fig. 6. Crayfish heart, Gold preparation of. (Acetic acid, 5 per cent., a few seconds, gold twenty minutes, formic acid one hour, at 40°C.) $\frac{1}{8}$ obj.

Fig. 7. Muscle-fibre of Daphnia. (Acetic acid, 5 per cent., ten minutes, gold thirty minutes, formic acid two hours, at 40°C.) $\frac{1}{10}$ imm. obj.

Fig. 8. Muscle of Bird (left in formic acid three days). $\frac{1}{10}$ imm. obj.

Fig. 9. Muscle-cell from Rat's heart, showing network. Usual gold method. Zeiss, D obj.

Fig. 10. Muscle-fibre from Frog's heart. (Osmic acid, 1 per cent., half an hour.) $\frac{1}{10}$ imm. obj.

Fig. 11. Network from heart-muscle of Bird. (5 per cent. acetic acid fifteen minutes, gold thirty minutes, formic acid twenty-four hours.) $\frac{1}{10}$ imm. obj.

Fig. 12. Portion of unstriped muscle-cell from mesentery of Newt, showing intranuclear network and its connection with the fibrils of the cell. (5 per cent. amm. chromate twenty-four hours; logwood.) $\frac{1}{10}$ imm. obj.

Fig. 13. Part of fibre from bladder of Salamander, showing fibrils. Usual gold method. $\frac{1}{10}$ imm. obj.

Fig. 14a. Muscle-fibre of Dytiscus, stimulated with alcohol vapour. Portion a relaxed. b. Contracted. $\frac{1}{10}$ obj.

Fig. 14b. Network of relaxed portion. $\frac{1}{10}$ imm. obj.

Fig. 14c. Network of contracted portion. $\frac{1}{10}$ imm. obj.

Fig. 15. Diagram of a muscle-fibre, showing change in network during contraction.

Fig. 16. Diagram of the intracellular network of striped muscle. a. The transverse networks. b. The longitudinal bars of the network. (Copied from Melland, loc. cit., Diag. 1.)

Fig. 17. Portion of network on a larger scale. (Copied from Melland, loc. cit., Diag. 2.)

Fig. 18. Portion of muscle-fibre of Dytiscus, showing the network very plainly. One of the transverse networks is split off, and some of the longitudinal bars are shown broken off. (Copied from Melland, loc. cit., Fig. 6.)

Fig. 19. Hypothetical diagram of the termination of nerve in muscle-fibre and the connection with the network; based on views discussed in the paper. S. Sheath of Schwann, continuous with sarcolemma. n. Axis-cylinder branching and connected with muscle-corpuscles. m. Muscle-corpuscles connected with transverse networks.

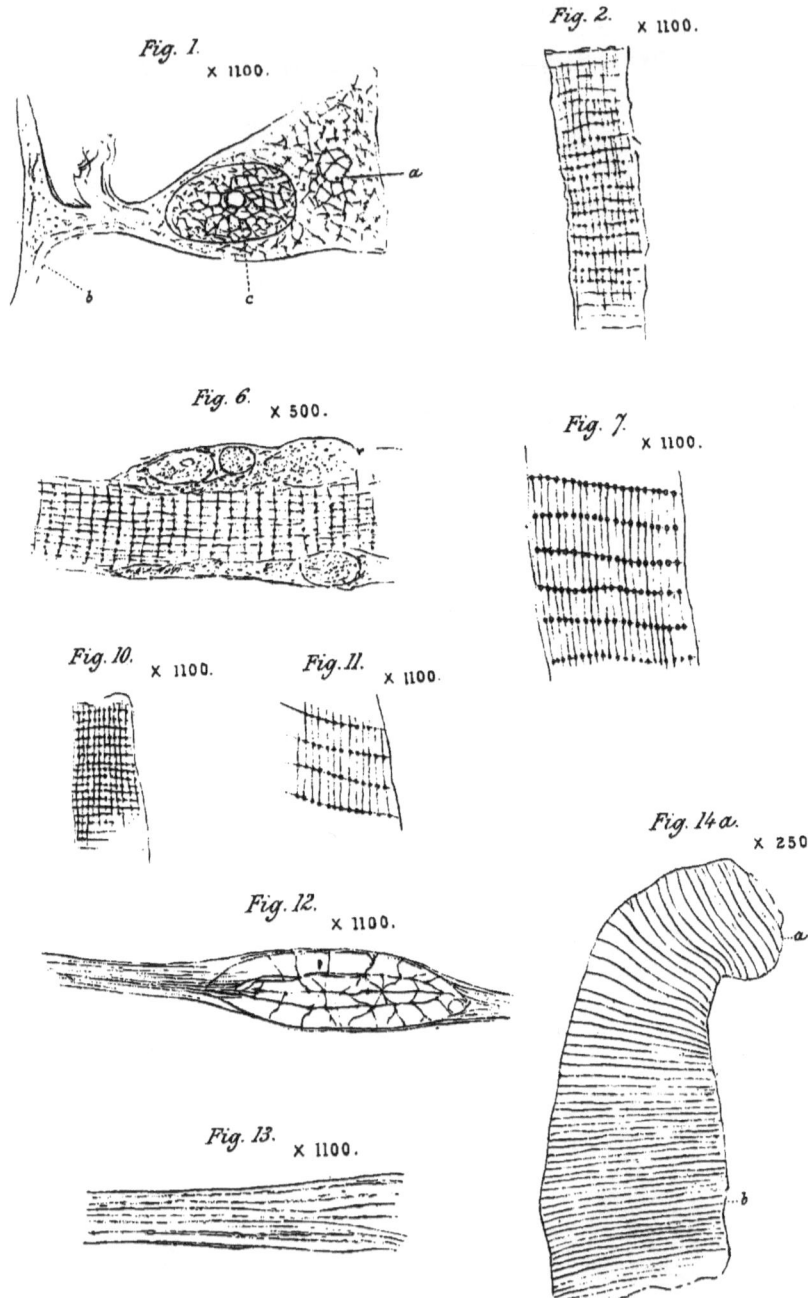
C. F. Marshall del.

Plate VI.

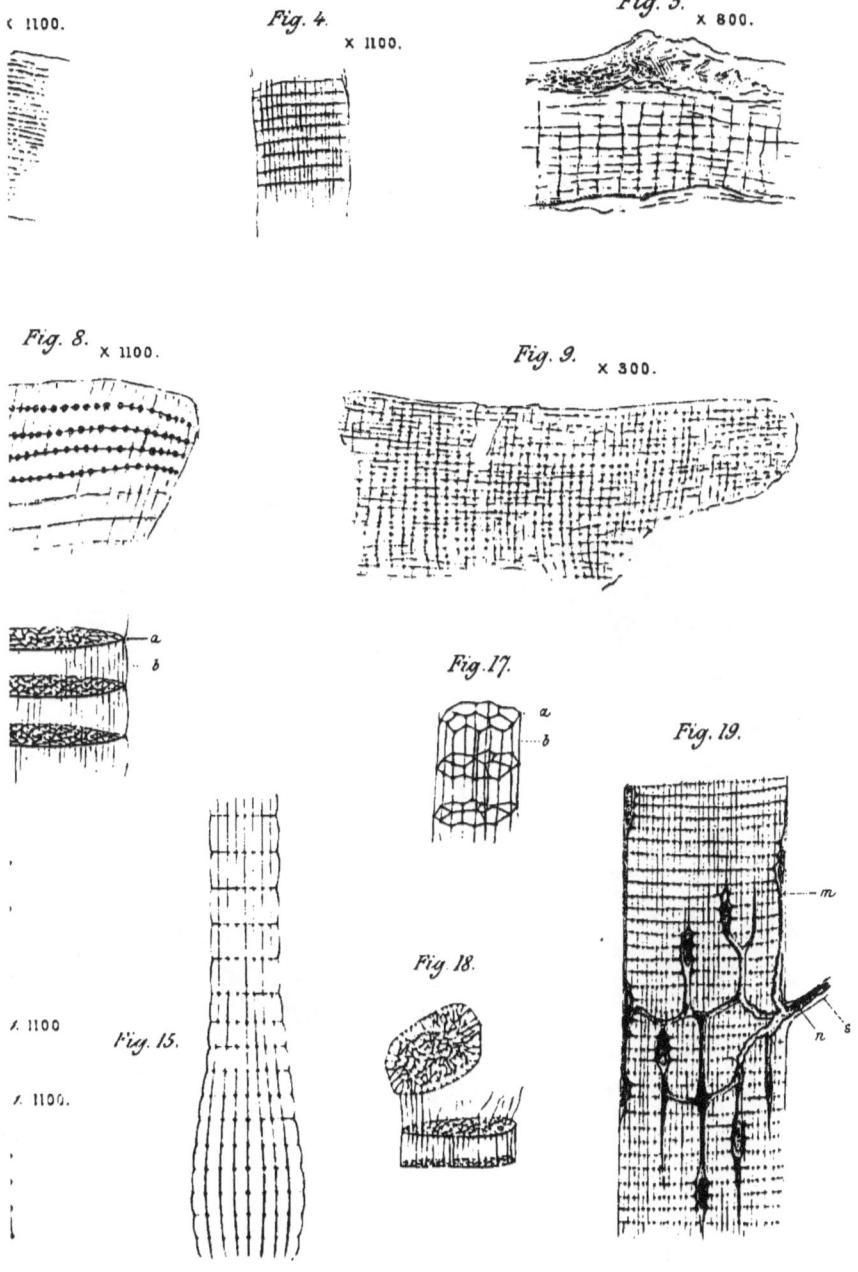

ON THE FATE OF THE MUSCLE-PLATE, AND THE DEVELOPMENT OF THE SPINAL NERVES AND LIMB PLEXUSES IN BIRDS AND MAMMALS.

BY A. M. PATERSON, M.D.,

Senior Demonstrator of Anatomy, and Lecturer in Dental Anatomy and Physiology, in the Owens College, Manchester.

[WITH PLATES VII. AND VIII.]

THE late Professor Balfour* showed that the spinal nerves in Elasmobranchs spring entirely from epiblastic origins, and the same has been proved conclusively regarding the roots at least of the spinal nerves in birds and mammals, by the researches of Milnes Marshall,† His,‡ and others. The most complete account of the early stages in the development of the nerves in higher Vertebrates is that of Marshall. He has traced the roots of the nerves from their origin from the spinal cord to the point when they unite together to form the mixed nerve. From that point onwards there is uncertainty. Though it is considered highly probable that the further growth of the nerves consists of an extension towards the periphery of the original epiblastic elements, still it has not been proved that this is so. It has not hitherto been shown that the nerve-trunks, after the junction of the two roots, are not formed from the cells of the mesoblast.

The present investigation has been undertaken with the object of tracing the nerves in their development from the condition in which

---

* "Monograph on the Development of Elasmobranch Fishes," London, 1878.

† "Journal of Anatomy and Physiology," vol. xi., p. 491.

‡ "Ueber d. Anfänge d. Peripherischen Nerven Systems," "Archiv f. Anat. u. Phys.," 1879.

previous embryologists have left them, to a point at which they may be fairly compared with the adult state.

Chick embryos, artificially incubated, have been used for the most part. By means of series of continuous sections, cut in different directions, the nerves have been traced from end to end in the different stages of development, from their first appearance up to the formation of trunks having an arrangement closely similar to that found in the adult. These sections have been compared with sections of mammalian embryos of different ages, with the result that the condition of development of the nerves has been found to be identical in both at periods in which the development of other parts and organs of the body is the same.

The methods adopted were in almost all cases the same. For hardening the Mammalian embryos, Kleinenberg's solution of picric acid was used; for the Chick embryos, a cold saturated solution of corrosive sublimate. A solution of borax carmine was the staining agent employed, prepared according to Balfour's directions. The sections were cut with a Jung's microtome, fitted with an ordinary razor.

The earliest stages in the development of the spinal nerves in the Chick have been described by Marshall. He has shown that they spring from the spinal cord as buds, of which the dorsal are the first to appear, arising from the summit of the cord. The more anterior of the dorsal roots arise from a "neural ridge," an elevation continued back from the hind brain. By the interstitial growth of the dorsal portion of the spinal cord these roots become separated, and their attachments to the cord more laterally placed. The ventral roots appear at a later date. Projecting outwards directly, they unite at an acute angle with the dorsal roots to form the mixed nerve. The spinal ganglion on the dorsal root is evident before the fusion of the two roots occurs; it is formed by the proliferation of the cells of the bud which form the root. The spinal nerves are thus formed in pairs, which occupy the intervals between the muscle-plates.

Before tracing the further growth of the nerves from this point, it is necessary to describe the destination of the muscle-plates and the mode of formation of the limbs, as in their onward development the nerves present differences according as they occur in relation to the limbs, or in the intervals between them.

I. Fate of the Muscle-Plate.

The dorsal and ventral roots of the nerves unite towards the end of the third day. But even before this time the limb buds have begun to appear (at sixty hours). The growth and differentiation of these buds, and the relations of the muscle-plates in the different regions of the body, complicate the process of nerve development.

In a Chick embryo at the age of three days, when transverse sections are made through the trunk between the limbs, the muscle-plate (*m. p.*, Fig. 1) is seen as an elongated column of cells lying directly beneath the epiblast, and separated from the spinal cord by the spinal ganglion (*sp. g.*) and roots of the nerve (*N.*). The lower end lies outside the angle (*a.*), between the somatopleure and splanchnopleure. The nerve-roots alternate with, and lie at a deeper level than, the muscle-plates. The muscle-plate itself consists of a double layer of cells, continuous at the ends, and separated from each other by a very evident line, the remains of the original cavity between the two strata. The outer layer, whose thickness is made up of several cells, consists of ovoid, spindle-shaped, or rounded cells, fitting closely together, and with their long axes directed from without inwards. They are sometimes multinucleated, and stain deeply with carmine. Round the ends of the muscle-plate they merge with the cells of the inner layer. The inner layer of cells has different characters. As seen in longitudinal sections, it is composed of spindle-shaped cells, which lie close together with their long axes directed from before backwards. Several of these cells occur in one somite in a line from front to back. In other words, the fibres are shorter than the thickness of the somite. In transverse sections the fibres appear rounded with large nuclei, and are more separated from one another. The cells at the upper and lower ends of the muscle-plate stain most deeply.

The bud which gives rise to the fore limb has at this date attained considerable size. It projects almost directly outwards from the side of the trunk (Fig. 2), growing partly from the mesoblast above the angle between somatopleure and splanchnopleure, and partly from the somatopleure itself. It consists of a mass of mesoblastic cells, densely packed together, especially at the surface and distal end.

Towards the origin of the limb the cells become more scattered. This mass of mesoblast is covered by a layer of epiblast, which presents two thickenings where the cells are in most active growth, one forming a cap covering the pointed outer end, the other forming a ring round the root of the limb, and appearing in transverse sections as thickenings above and below the root of the limb. The mesoblastic cells are entirely undifferentiated as yet; they are rounded, and stain deeply. Embryonic blood spaces are found here and there. The position of the muscle-plate is different from what has been described in the region between the limbs. By the growth of the limb bud it has become separated from the somato-splanchnopleuric angle (*a.*). Its lower end reaches to about the centre of the upper half of the base of the limb, that part which is continuous with the intermediate cell mass. Moreover, the plate does not lie external to the angle of the body cavity; otherwise it has the same relative position as in the dorsal region to the parts which constitute the trunk at this period. Histologically, also, it has very similar characters. The differences are seen in the centre of the plate. The inner stratum of longitudinally arranged spindle-cells is thicker. It stains badly, and the nuclei of the fibres are large and well defined. The outer layer is thinner in the centre, becoming thicker when traced towards the end of the plate. The cells of this layer stain deeply. They are scattered in the centre, becoming more closely packed at each end.

In Chick embryos at three days six hours (Fig. 3), in the dorsal region the muscle-plate (*m. p.*) has extended a short distance down the body wall in the somatopleure. The central portion of the plate is thicker, owing to an accumulation of the longitudinal fibres of the inner layer. The cells of the outer stratum are still more separated in the centre, and do not stain so deeply. The layer as a whole is thinner. The ends of the plate still present the primitive condition of the cells, which are angular, packed close together, and stained deeply. In the region of the fore limb at this date (Fig. 4) the muscle-plate (*m. p.*) has the same characters; but its position is different. Its lower end reaches no farther than midway between the dorsal attachment of the limb to the trunk, and the somato-splanchnopleuric angle.

Twelve hours later (at three days eighteen hours) these changes are seen to be more pronounced. In the dorsal region (Fig. 5), owing partly to the increase in vertical extent of the embryo, the muscle-plate (*m. p.*) has become elongated, and at the same time thinned out. There is no distinct trace of an outer layer to be found, except at the ends of the plate. Here the cells retain their primitive character, and stain deeply. The rest of the plate consists entirely of longitudinally arranged fusiform cells. The muscle-plate has a peculiar bend, passing almost vertically downwards towards the body cavity, and then suddenly sweeping outwards to enter the body wall. Its lower end has passed still farther down the somatopleure, lying close to the inner side. In the fore limb at this date (Fig. 6) the muscle-plate retains its original position. It does not extend outwards farther than the somato-splanchnopleuric angle. Its histological characters are the same as in the dorsal region. The limb bud itself has increased in size, and is now directed downwards. The cells which compose it are still undifferentiated.

In embryos at four days, in the regions of the trunk between the limbs (Fig. 7), the muscle-plate (*m. p.*) has extended down through a third of the length of the body wall, lying close to the outer side of the body cavity. Its relations and structure are the same as before. Each end is surmounted by a cap of cells, which retain their primitive characters; rounded, fusiform, or angular, they stain deeply, and are plainly separated off from the main part of the muscle-plate. These cells can be traced on to the outer surface of the muscle-plate, where they gradually become lost. The main part of the plate consists of elongated fusiform cells. In the region of the fore limbs (Fig. 8) the mesoblastic tissue of the limb still presents the same characters. The cells stain deeply, are round, ovoid, and often multinucleated; but still undifferentiated. The fœtal vessels are better marked. The muscle-plate (*m. p.*) occupies its primitive position, ending below at the root of the limb. It has the same structural characters as in the dorsal region; but the undifferentiated cells, which stain deeply, are best marked at the upper end of the plate.

In embryos at four days twelve hours, in the trunk between the limbs, the muscle-plate (Fig. 11, *m. p.*) has passed half way down the body wall, lying close outside the cavity. It now consists almost entirely of elongated spindle-cells. In the region of the limbs further changes are taking place, but in which the muscle-plate takes no part. It remains within the body cavity, and ends below at the same limit as before (Fig. 12, *m. p.*). Structurally it is the same. In the limb buds themselves the first changes occur at this date, in the formative blastema, which result in the production of the muscular and osseous systems. The nerve plexuses, as we shall see below, have been produced; and the resulting trunks have passed into the limb, in two groups, one dorsal (*d.*), the other ventral (*v.*). Between, above, and below the nerves, the mesoblastic cells are taking on a characteristic arrangement. The cells immediately above and below each trunk (1, 2) are more closely packed together, forming thick layers, each several cells deep. They are histologically the same as before. In the centre of the limb bud (3), between the nerve trunks, the cells are now arranged in a concentric and symmetrical fashion, and are separated from one another by a small amount of intercellular substance. Towards the dorsal and ventral surfaces of the limbs the cells are more scattered, so that the central portion of the limb in transverse sections appears darker than the superficial parts. At the free end of the limb (4) there is still no differentiation of tissue elements. The cells here form a simple mass, without any distinction into layers or groups.

In a five days' embryo the condition of the muscle-plate in the region between the limbs (Fig. 13, *m. p.*) is much the same as at the last mentioned period. It shows still further extension in a ventral direction down the body wall. In the same embryo, in the regions of the limbs the muscle-plate (Fig. 14, *m. p.*) has clearly no connection with the muscular system of the limb itself. It consists now of elongated fibres, forming, in transverse sections, a column lying just outside the spinal cord and nerves, and separated from the surface of the body and from the limbs by a considerable thickness of ordinary blastema. The several tissues of the limb are formed from mesoblastic elements, developed *in situ*. A central core of cellular cartilage (3) is very evident at this date. The cells are arranged regularly in a concentric manner in transverse sections; in transverse rows in longitudinal sections. This cartilaginous cylinder is found in longitudinal sections to be broken up into segments, corresponding to the skeletal elements of the limb. The intercellular substance between the cells has largely increased in amount. At the periphery of this central mass the cells gradually become changed in character, being more deeply stained in mass, and individually becoming oat-shaped or fusiform. The long axes of the cells are directed from within outwards. Two layers of cells thus appear, one above, the other below the cartilaginous bar. The cartilage, however, is not yet distinctly marked off, but is connected to these groups of cells by a definite and still more deeply-stained (perichondrial) layer. The nerves to the limbs derived from the plexuses have a very definite relation to these central groups of cells, which are enclosed within the nerve trunks. The dorsal nerves pass over, and the ventral trunks beneath, this central area. Above and below the nerves two other more distinct groups of ovoid cells (1 and 2) are now apparent, collected each into more or less separate subsidiary bundles, and easily distinguished from the surrounding undifferentiated mesoblast by their shape, and by the fact that, being closely packed together, they stain more deeply *en masse*. The four simple, unsegmented strata of ovoid cells—of which two are dorsal, and lie above and below the dorsal branches of the nerves; two ventral, and having a similar relation to the ventral trunks—are evidently the precursors of the muscular elements of the limbs. They are quite distinct from the muscle-plates, and are separated from them by a large quantity of undifferentiated mesoblast. The blood-vessels of the limb have now attained a large size, and are regular in their arrangement. One large artery (*art.*) passes down the centre of the limb on its ventral aspect, lying among the nerves, and accompanied by a vein (*V.*).

It is unnecessary to follow the muscle-plate further. It has been shown that while it grows into the body wall between the limbs, and forms the basis of the longitudinal muscles of the trunk; in the region of the limbs, it remains in its primitive position, and has no share in the formation of the limb-muscles. It is merely concerned here in forming the longitudinal muscles of the back. In the limbs themselves the muscles are produced by the further differentiation of the four dorsal and ventral strata, which have been described as appearing from the mesoblast cells, at first undifferentiated, and forming the original limb bud. In Chick embryos at six days, these strata of ovoid cells have become more fusiform, and are collected into more definite and separate systems. Two days later the muscles throughout the body are seen distinctly.

II. Development of the Spinal Nerves and Limb Plexuses.

The later development of the spinal nerves naturally divides itself into two parts: firstly, in relation to the limbs; and secondly, in the trunk between the limbs. In essential points the processes are the same in all regions of the body. The formation of the limbs, however, and the peculiarities in the position of the muscle-plates, give rise to certain differentiations in the arrangement of the nerves in those regions.

In Chick embryos at three days (Figs. 1 and 2), both in the trunk between the limbs and in the regions where the limbs are being formed, the nerves have reached the same stage of development. The nerve roots, which lie within and alternate with the muscle-plate, have joined together. From their fusion a slender, finger-like process of cells results (N.), which represents the commencing nerve trunk. The dorsal root is oval in transverse section, the ganglion, which is very large, forming nearly the whole of it. The nerves and their roots consist of large ovoid cells, containing often two or three nuclei, the long axes of the cells being directed outwards from the cord. They stain more deeply than the mesoblast cells in which they lie, and are surrounded by a slight amount of feebly-stained intercellular substance.

Six hours later (at three days six hours), the slender stalk (N., Figs. 3 and 4), retaining the same position within the muscle-plate, has grown downwards and outwards as far as the somato-splanchno-pleuric angle (a.). It has the same relative position in the trunk and in the regions of the limbs, passing between the muscle-plate and cardinal vein ($c.\ v.$). But, owing to the difference of growth of the muscle-plates in the two regions, it has reached its lower end in the regions of the limbs (Fig. 4); while in the trunk, the muscle-plate, having by this time entered the body wall (Fig. 3), extends farther than the nerve.

This description corresponds with the condition of development of the spinal nerves and muscle-plates in the Rabbit embryo of seven or eight days. Both histologically and morphologically the nerves have reached the same state of development.

In Chick embryos of three days eighteen hours there is not much difference in the relative amount of the growth of the nerves. The whole embryo has, however, increased in size. In the trunk (Fig. 5) the nerves cannot yet be traced into the body wall. In the region of the limbs (*N.*, Fig. 6) they have passed out beyond the lower end of the muscle-plate and beyond the angle of the body cavity. The marked change, however, at this date, is in the histological structure of the nerves. The spinal ganglia are well-formed ovoid masses, composed of large ovoid cells with two nuclei, the cells having a general arrangement in vertical rows. The cells forming the nerve-trunks have become elongated, fusiform, with fibrillar processes at each end. The body of the cell does not stain well; the nucleus, large, oval, and with several nucleoli, lies in the centre of the cell, and stains deeply; the distal ends of the nerves in the regions of the limbs present a ragged appearance, due to the protrusion and separation of these spindle-shaped cells into the mesoblastic tissue.

At four days the histological change is more marked. The cells forming the nerve-trunks have become more fibrous, the nuclei are less numerous, and the trunks stain yellow *en masse*. Both between the limbs and in relation to the limb buds the growth of the nerves has continued. In relation to the limbs (*N.*, Fig. 8), the nerves sweep round between the lower ends of the muscle-plates and the body cavity, and, reaching the base of the limb, spread out, and then divide into a sheaf of branches, which diverge in the formative blastema of the limb. In the trunk between the limbs the nerves (*N.*, Fig. 7) have extended a great way down the body wall, lying between the muscle-plate and the body cavity, but not reaching as far as the lower end of the plate. They divide, as in the limb, into branching processes, some of which pass directly outwards into the muscle-plate and divide again; some pass on, lying within the muscle-plate.

It is at this period that I have first been able to make out satisfactorily the existence of the trunk passing to join the sympathetic. A slender cord arises from the spinal nerve midway between the junction of the roots and the distal end. It courses inwards at right angles to the main trunk, and is soon lost. Now also the formation of the superior primary division of the nerve is first seen distinctly. It is constructed in the same way in mammals, and is seen still more clearly in Rat embryos at fifteen to seventeen days (Fig. 16).

Each root of the nerves divides into two unequal branches—the dorsal root beyond the ganglion, the ventral root directly. Of these branches the smaller is superior, the larger inferior in both cases. The larger branches unite to form the main trunk of the nerve, or the inferior primary division; the smaller branches combine to form the superior primary division. This is directed upwards and outwards, and sub-divides as it passes towards the surface.

In Chicks at four days six hours, the condition of the nerves in the trunk between the limbs is slightly more advanced, but presents no change of any note. In the regions of the limbs, however, the plexuses are now formed. In transverse sections through the embryo, the nerves are found, on entering the limbs, to divide into two fairly well-defined strands, separated by a central mass of mesoblast, which is in active growth, and preparing to form the cartilaginous basis of the limb. The nerves, in fact, spread out around this central core, and arrange themselves into two sets, one dorsal, the other ventral. These dorsal and ventral branches of the nerves only pass a short distance into the limbs, and are not so well defined as in embryos a few hours older. But even now the process of plexus formation may be seen.

When longitudinal vertical (sagittal) sections are made continuously through the body, it is seen that the nerves to the limbs, besides forming the dorsal and ventral branches above mentioned, unite with adjacent nerves at the root of the limb to form a well-defined plexus. In the Fowl three main trunks form the brachial plexus,* the first thoracic and the last two cervical nerves, with, in addition, a small branch from the more anterior cervical nerve. When sagittal sections are made, the limbs being divided transversely at their roots (Fig. 9, *a.*), the axillary artery (*art.*) and vein (*v.*), with the three main nerves (*N.*, 1, 2, 3), are divided just outside the body cavity (*B. C.*), and as they lie in the body wall (*B. W.*), before their entrance into the limb bud, and below the terminations of the muscle-plates. In successive sections from within outwards, these nerves can be traced to their terminations in the limbs. They first spread out, and approach one another, as described; in doing so they unite with adjacent nerves, so that the next step in the pro-

---

* Macartney, Art. "Birds" ("Rees' Cyclopædia").

ceeding (Fig. 9, *b.*) is the formation of a plexiform mass of nerve-tissue (*plex.*), which encircles the artery. At the same time that this plexus formation occurs, the division into dorsal and ventral branches is beginning. In the figure the axillary artery in the centre, with a group of mesoblast cells running up and down from it, shows the commencing separation of the mass into these two portions. In sections made a little farther outwards, the plexus gradually separates completely (Fig. 9, *c.*) into a dorsal and a ventral mass (*d.* and *v.*), each consisting of a broad, flattened band, separated by the artery and some mesoblastic cells. Still farther out, just before the limb is completely separated from the trunk (Fig. 9, *d.*), the nerves appear as two distinct cords (*d.* and *v.*). These gradually divide, become attenuated, and disappear as they are traced towards the distal portion of the limb. Exactly the same process occurs in relation to the nerves of the hind limb. When oblique longitudinal sections are made at this date, so as to cut through the length of the nerves (Fig. 10), they are seen to spread out and divide laterally, so as to unite with similar branches, dorsal or ventral, of adjacent nerves, to form the plexus at the root of the limb. As already stated, there is no trace whatever at this date of either cartilage or muscle in the limb, which consists entirely of undifferentiated blastema.

In embryos four days twelve hours old the nerves have reached a more advanced stage of development. In the trunk between the limbs (Fig. 11, *N.*) the nerve, which twelve hours earlier divided into a sheaf of branches in the body wall, now splits into two well-defined and unequal branches at a point just beyond the somato-splanchnopleuric angle. The larger branch continues the direction of the main nerve, and can be traced for some little distance between the muscle-plate and body cavity. The smaller nerve represents the lateral branch of the adult, and is directed downwards and outwards through the muscle-plate, on the outer side of which it divides and is finally lost. The cord to the sympathetic nerve can be followed farther than before; but I have been unable to trace its connections with the roots and trunk of the spinal nerves.

In the limbs the early changes occurring in the blastema, which lead to the production of the osseous and muscular systems, have already been described. The nerves are of large size, and can be traced more than half-way through the limb towards the distal end. At the root of the limb the main trunk (Fig. 12, *N.*) can be seen in transverse sections to divide into two well-defined trunks, which are clearly homologous with the terminal branches of the nerve in the region of the trunk at this date (Fig. 11). These two large nerves are respectively dorsal (*d.*) and ventral (*v.*); and enclose between them the densest portion of the blastema (3). This enclosed portion has already been described as consisting of three parts—a central part, which is going to form cartilage, and a dorsal and ventral part, the elements of muscular tissue. On the dorsal surface of the dorsal nerve, and on the ventral surface of the ventral nerve, are other layers of mesoblast cells (1 and 2) undergoing division preparatory to the production of muscles. In longitudinal sections the three main nerves supplying the fore limb can be traced as before in successive sections, each dividing into dorsal and ventral branches; and these again uniting with adjacent nerves to form two flattened bands, which pass to the dorsal and ventral surfaces of the limb.

The nerve-trunks are almost entirely fibrous now, with rows of deeply-stained nuclei arranged among the wavy fibres. These are evidently the connective-tissue elements of the nerve-trunks. Towards their terminations the fibres are fewer and the fusiform nerve-cells more abundant.

In Rat embryos between twelve and fourteen days old, exactly the same condition of development of the nerves is found as in the Chick at four days twelve hours. The state of development of the body generally is the same, the limbs exist as buds projecting downwards and outwards from the trunk, and composed, for the most part, of undifferentiated blastema. In the centre of this the cells are more closely massed together than at the periphery, and are being arranged concentrically to form the cartilaginous basis of the limb. Each of the roots of the nerve divides into upper and lower branches, which respectively unite; the upper branches to form the superior primary division, the lower to form the inferior primary division of the nerve. The latter is the main trunk. Passing downwards and outwards below the muscle-plates, it reaches the base of the limb, where it divides into two branches, one dorsal, the other ventral, with regard to the cartilaginous core. These branches can be followed through the limb, actually as far as the epiblastic surfaces and almost to the distal end. In minute structure the trunks very much resemble the nerves in the Chick, the chief difference being that the fusiform cell elements are more evident throughout.

In Chick embryos at five days the changes in the nerves between the limbs are not marked (Fig. 13, N.). The lateral and inferior branches are well defined, and the whole nerve has passed farther down the body wall. From this time onwards these trunk-nerves present no marked differences in morphological arrangement from what is found in the adult.

In the regions of the limbs at this date, as already described, the cartilaginous basis and muscular elements have begun to make their appearance. The nerves themselves occupy a position with regard to these elements which is highly characteristic. The dorsal and ventral trunks (Fig. 14, $d.$ and $v.$) are each covered above and below by masses of specialised, oat-shaped cells, which represent the layers of dorsal and ventral muscles. These double dorsal and ventral muscular layers are also separated by the cartilaginous framework of the limb. The nerves themselves stain yellow; consist of extremely wavy fibres, and present no distinct nuclei. Deeply-stained connective-tissue corpuscles lie among the fibres.

At five days twelve hours the process of muscle and cartilage formation in the limbs is more advanced. The appearance of the nerves in transverse section is much the same as before. In successive longitudinal (sagittal) sections (Fig. 15) the nerves can be seen at the root of the fore limb, undergoing division and union in the brachial plexus in the same way as, but more definitely than in younger embryos (four days six hours, Fig. 9). The three main trunks are seen first (Fig. 15, $a.$, $N.$, 1, 2, 3) in company with the axillary artery and vein. They then divide, in successive sections (Figs. 15, $b.$—15, $c.$), into dorsal ($d.$) and ventral ($v.$) branches. The dorsal branches unite with dorsal branches, the ventral branches with ventral branches, to form the nerves of distribution to the limbs. The plexus formation is now complete, and from this time onwards there is no change in the *essentials* of its formation, which tally with the condition of the adult brachial plexus. It is to be borne in mind that, though the plexus is completely formed, yet the muscular elements are in a simple condition. Muscles are not formed in anything approaching the complexity of the adult limbs sooner than the ninth day.

III. Conclusions.

1. *On the fate of the muscle-plate and development of the muscles of the limbs.*

In Elasmobranchs the muscles of the limbs are formed by the muscle-plates which grow down, and as they pass the roots of the limbs give off small buds, which become separated off and form the starting point of the muscle formation in the limbs.* In higher Vertebrates, while it has been held probable† that the muscle-plates do not enter into the formation of these muscles, it has not been shown satisfactorily how they do arise, and what becomes of the muscle-plates. It is evident from the foregoing details that the muscle-plates, in the regions of the limbs, stop short in their downward growth, do not pass farther than the base of the limb, and are not concerned in any way with the production of the limb muscles.

These are formed by the differentiation of the mesoblastic cells forming the primitive limb buds. These cells form, in the first place, a central cartilaginous bar, above and below which, in the second place, are developed a double dorsal and a double ventral layer of simple muscle, which later on becomes more complex in its arrangement, and forms the muscles of the adult.

In the region of the trunk, between the limbs, a different disposition of the muscle-plates occurs. They grow down in the body wall, and eventually become converted into the longitudinal muscles of the trunk. They do not, however, appear to assist in the formation of the sub-vertebral (hyposkeletal) muscles.

In both regions the growth and differentiation of the parts of the muscle-plates are alike. The outer layer disappears gradually; the inner layer of the plate being converted into the longitudinal fibres. The disappearance of the outer layer is possibly due to the conversion of the cells into longitudinal fibres, which merge with those of the inner layer; but this is not certain. In Elasmobranchs each

---

* Balfour, "A Monograph on the Development of Elasmobranch Fishes," London, 1878.

† Kölliker, "Entwickelungsgeschichte d. Menschen u. der höheren Thiere," Leipzig, 1879.

fusiform cell extends from end to end of the muscle-plate.* In birds and mammals this is not so. Each fibre is considerably shorter than the breadth of the somite.

The chief point of interest here is in connection with the development of the limb muscles. They first appear as double layers of dorsal and ventral cells, which layers are simple, without segmentation, and derived from the mesoblast cells of the primitive limb bud. In Elasmobranchs† this stage in the development of the limb muscles is a secondary one, and is preceded by events which are omitted in higher Vertebrates. The process of evolution of the limbs in birds and mammals is therefore shortened. In Elasmobranchs a downward growth and a cutting off of part of certain muscle-plates occur; the portions cut off undergo further growth, passing into the limb bud, fusing together, and becoming differentiated into dorsal and ventral strata. In birds and mammals the same end is reached without these preliminary steps, and without the intervention of the muscle-plates. The definite relations which these simple muscular layers bear to the nerves of the limbs, throw light on the evolution of the limb plexuses. Each nerve, passing into a particular region of the limb bud, divides into dorsal and ventral branches, to supply the dorsal and ventral surfaces respectively of that particular portion. As the mesoblast forming the limb bud becomes more differentiated, so as to give rise to the muscular layers, the portions opposite to and originally derived from the same somites as the nerves become fused, forming simple muscular layers in the first place. The nerves therefore fuse together; the dorsal branches forming a dorsal band, and the ventral branches a ventral band, which pass out, and are finally lost in these simple muscular layers.

2. *On the growth and development of the spinal nerves.*

It is difficult to demonstrate clearly, but it is next to impossible to deny, that the spinal nerves are developed from epiblast throughout their whole length. From the numerous sections which I have examined at different periods of growth, I have traced the spinal nerves, not only the nerve-roots, but also the trunks and the plexuses, as a centrifugal growth from the spinal cord. The growth

---

* Balfour, "Comparative Embryology," p. 552.
† Balfour, "Monograph on Elasmobranchs."

of the nerves is both interstitial and terminal. Consisting at first merely of rounded cells, in an active state of proliferation; in older embryos these become first ovoid and then fusiform, at the same time being less deeply stained with borax carmine. These fusiform cells, by the alteration of their protoplasm, become converted into nerve-fibres. Moreover, while this interstitial growth goes on, the trunk of the nerve is elongated by means of proliferation of the cells at the periphery, which retain a primitive character longer than those in the more proximal portion of the trunk. For example, when the cells are fusiform in the nerve near the cord, they are oat-shaped at the distal end; when they are fusiform at the distal end of the nerve, they are fibrous in the proximal part of the trunk.

3. *On the homologies of the spinal nerves.*

Their development shows that the nerves which form the limb plexuses are homologous with the whole nerves in the regions between the limbs, where their arrangement is simplest, and not merely with the lateral branch, as Goodsir supposed.* The nerves in both regions first spread out into a ragged bundle. These bundles at a later period arrange themselves into two well-defined cords, the division from the main trunk having the same relative position in both. In the regions between the limbs these trunks represent the lateral and inferior branches; in the regions of the limbs they are dorsal and ventral in position.

4. *On the development of the limb plexuses.*

I have elsewhere shown† that in mammals, and as far as I have been able to make out, the same holds good for birds also, the limb plexuses are formed on a definite plan, which is essentially the same in all the animals examined, and in relation to both fore and hind limbs. The nerves which form the plexus divide, in the first place, into dorsal and ventral branches. These divisions sub-divide, and the secondary cords, whether dorsal or ventral, combine with the cords formed by the division of adjacent (dorsal or ventral) trunks to form the nerves of distribution. Any given nerve to the limb

---

* "Edinburgh New Philosophical Journal," New Series, vol. v., Jan., 1857; "Anatomical Memoirs," vol. ii., p. 201, 1868.

† Graduation Thesis, Univ. Edin., 1886, "On the Spinal Nervous System in Mammalia;" "The Limb Plexuses of Mammals," "Journal of Anatomy and Physiology," vol. xxi., 1887, p. 611.

may be derived from any number of the spinal nerves constituting the plexus, but it is always formed by a combination of either dorsal or ventral nerves.

This mode of arrangement of the nerves in the plexuses is to be explained by a reference to their embryology, and the mode of development of the different parts of the limbs. The plexus formation is complete, and the nerves of distribution are formed in the embryonic limb long before the appearance of muscles. In the development of the nerves in the limbs the following steps occur. The primitive nerve, in the first place, grows out beyond the lower end of the muscle-plate, and reaches the root of the limb. It there, secondly, spreads out into an irregular series of processes, which pass into the undifferentiated tissue of the limb. Thirdly, these branches, at a later date, arrange themselves in two trunks, one dorsal, the other ventral, which extend still farther into the limb and enclose between them a mass of blastema, from which the cartilaginous basis of the limb is formed. Fourthly, the dorsal and ventral trunks fuse with adjacent dorsal and ventral trunks to form two broad flat bands, from which, still later, the individual nerves as found in the adult are produced.

The development of the muscular system of the limb, occurring after the formation of the nerves, corresponds with it exactly. The muscles appear first as simple double dorsal and ventral layers, among which the nerves pass as dorsal and ventral bands, formed by the fusion of adjacent dorsal or ventral divisions of the nerves of origin. As these muscular strata lose their simplicity and take on the complex arrangement of the adult, the nerves at the same time become more and more sub-divided, until in the adult the primitive characters of both are considerably masked.

Still, in the adult mammal, it is evident that the more preaxial nerves in the series supply the more preaxial portions, the postaxial nerves the postaxial portions of the limb,[*] and the combinations of dorsal divisions and ventral divisions of the nerves are distributed to those muscular and cutaneous areas which are derived respectively from the primitive dorsal and ventral surfaces of the embryonic limb bud.[†]

[*] Herringham, "On the Human Brachial Plexus," "Proc. Roy. Soc.," Jan., 1887
[†] "Journal of Anatomy and Physiology," vol. xxi., July, 1887.

In conclusion, I wish to express my deep indebtedness to Professor Milnes Marshall, for much advice and encouragement during the prosecution of the above researches, and for his kind assistance in the preparation of the present memoir.

EXPLANATION OF PLATES VII. AND VIII.

Illustrating Dr. Paterson's paper "On the Fate of the Muscle-Plate, and the Development of the Spinal Nerves and Limb Plexuses in Birds and Mammals."

Fig. 1. Semi-diagrammatic view of transverse section through the trunk of a Chick embryo at the end of the third day. Both spinal nerve ($N.$), with its roots and ganglion ($Sp. g.$) and muscle-plate ($m. p.$) are shown. The spinal cord, notochord ($No.$), aorta ($Ao.$), and cardinal vein ($C.V.$) are also indicated. The muscle-plate is just entering the body wall. The section is taken between the limbs.

Fig. 2. View showing the same structures in an embryo of three days in the region of the fore limb.

Fig. 3. From an embryo of three days six hours old, showing the growth of the muscle-plate ($m. p.$) and spinal nerve ($N.$) in the trunk between the limbs.

Fig. 4. From the same embryo in the region of the fore limb.

Fig. 5. From an embryo aged three days fifteen hours, showing the further growth of the muscle-plate ($m. p.$) and nerve ($N.$) in the trunk between the limbs.

Fig. 6. Shows the same structures in the region of the fore limb.

Fig. 7. From an embryo aged four days, with the muscle-plate presenting growing points at the two ends, and the nerve dividing in the body wall into a ragged bundle. The formation of the superior primary division ($s.$) of the nerve and the cord to join the sympathetic ($sy.$) are also seen.

Fig. 8. From the same embryo, through the fore limb, showing the relative position of the muscle-plate and nerve. The nerve is seen dividing in the limb into a sheaf of branches.

Fig. 9. *a.* Longitudinal section through the body of a Chick embryo aged four days six hours, in the region of the fore limb, cutting through the body wall (*B.W.*) just below the level of the muscle-plate. The body cavity (*B.C.*) is seen, and in the body wall are the three nerves (*N.*, 1, 2, 3), and the artery (*art.*) going to the limb. *b.* This section is made farther out at the root of the limb bud, and shows the thickening of the body wall, with the formation of the plexus, around the artery (*art.*). The vein (*v.*) is also seen. The nerves in the plexus are here on the point of separating into two bundles. *c.* From a section made still farther out in the limb. The nerves, after forming the plexus, have more completely separated into a dorsal and a ventral bundle (*d.* and *v.*), with the artery in the middle. *d.* Here the limb bud is more apparent; it is becoming separated from the body wall; and the dorsal and ventral bands of nerve are well defined in the blastema.

Fig. 10. From a longitudinal section through the body of the same embryo in the region of the hind limb (*H.L.*), to show the lateral division and union of the dorsal and ventral divisions of the nerves to form the plexus (*plex.*). The spinal cord (*Sp. C.*) and spinal ganglia (*Sp. g.*) are also shown.

Fig. 11. Transverse section through trunk of Chick embryo aged four days twelve hours, between the limbs, to show the muscle-plate and the spinal nerve. The division of the latter into its two terminal branches (*l. i.*) is seen.

Fig. 12. Transverse section through the same embryo, in the region of the fore limb. The spinal nerve is seen dividing at the root of the limb into dorsal (*d.*) and ventral (*v.*) branches. These enclose and are surrounded by masses of formative blastema (1, 2, 3), the precursors of the muscles and skeleton of the limb. The muscle-plate (*m. p.*) occupies its original position.

Fig. 13. Transverse section through the trunk, between the limbs, of an embryo aged five days, to show the relative position and growth of the muscle-plate (*m. p.*) and nerve (*N.*).

Fig. 14. Transverse section through the trunk and root of the fore limb of an embryo of the same age. *m. p.* Muscle-plate. *N.* Spinal nerve. *d.* and *v.* dorsal and ventral divisions. 3. Commencing formation of cartilage in centre of limb, with a layer of dense blastema surrounding it. 2 and 1. Commencing formation of muscles above and below the nerve-trunks. *art.* Axillary artery. *v.* Axillary vein.

Fig. 15. Successive longitudinal sections through the root of the fore limb of a Chick embryo aged five days twelve hours. The defined borders represent the body cavity. From *a.* to *e.* the nerves (*N.*, 1, 2, 3) are seen to divide into dorsal and ventral branches (*d. v.*), which unite laterally to produce secondary dorsal and ventral trunks. *Art.* and *v.* Axillary artery and vein.

Fig. 16. Diagram to illustrate the formation of the superior (*S.*) and inferior (*I.*) primary divisions of a spinal nerve from a Rat embryo. *Sp. C.* Spinal cord. *Sp. g.* Spinal ganglion. *No.* Notochord. *Ao.* Aorta.

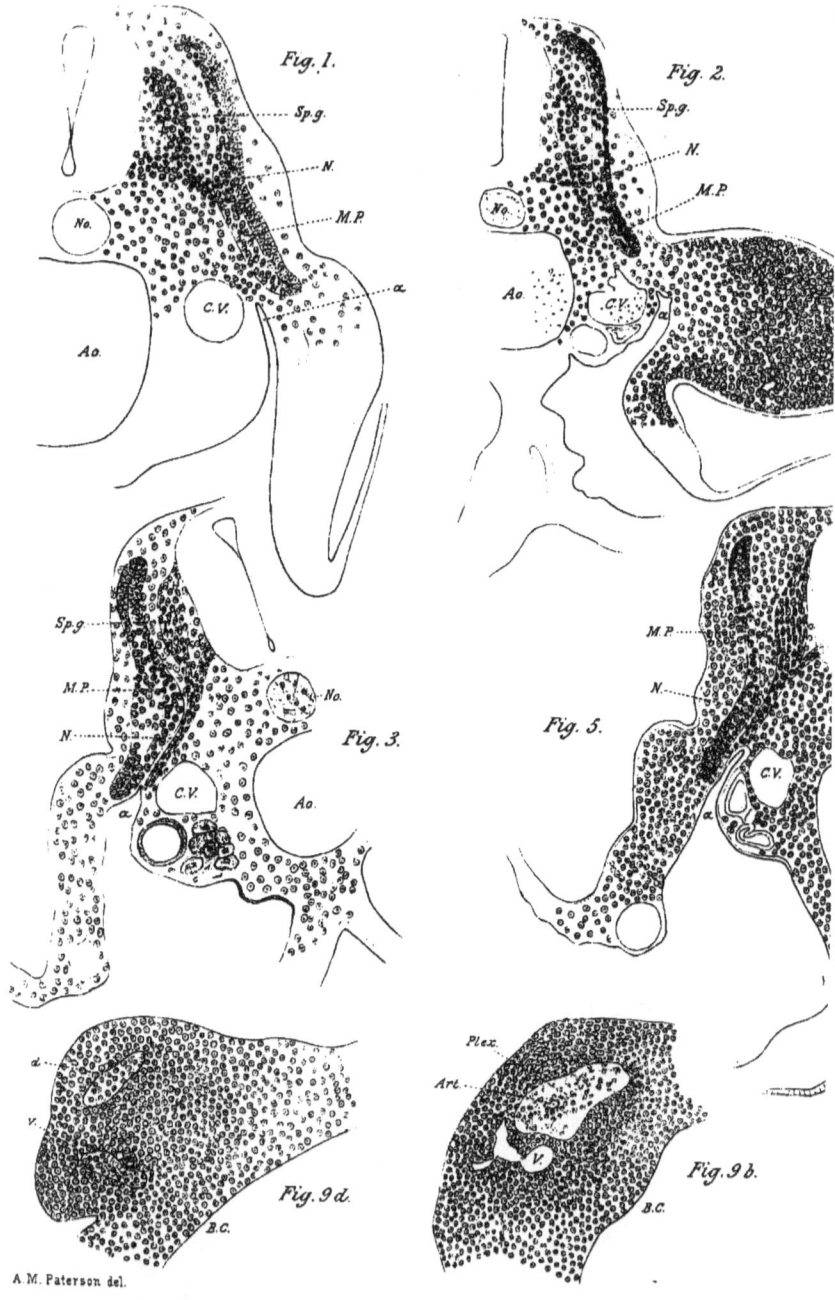

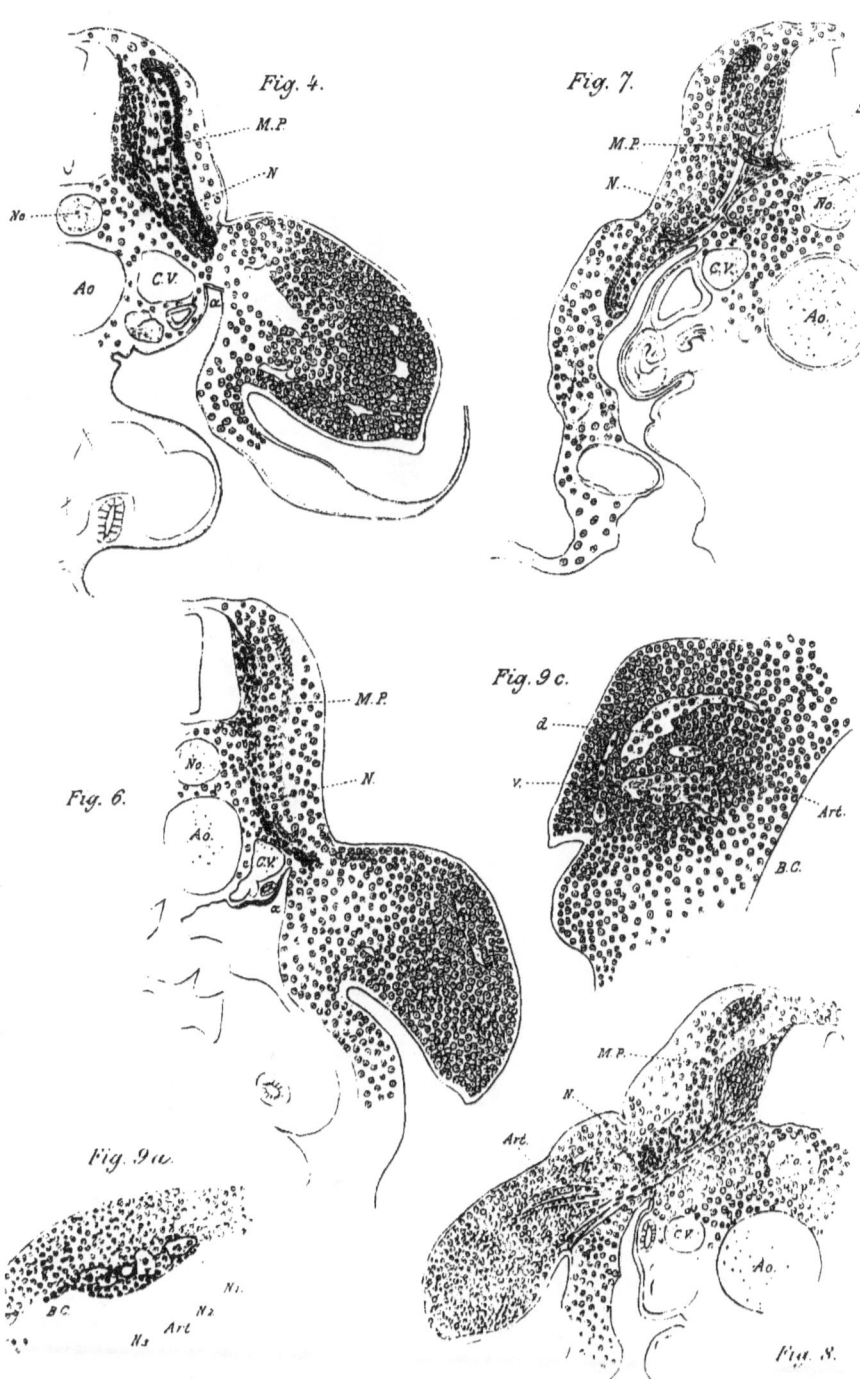

Plate VII.

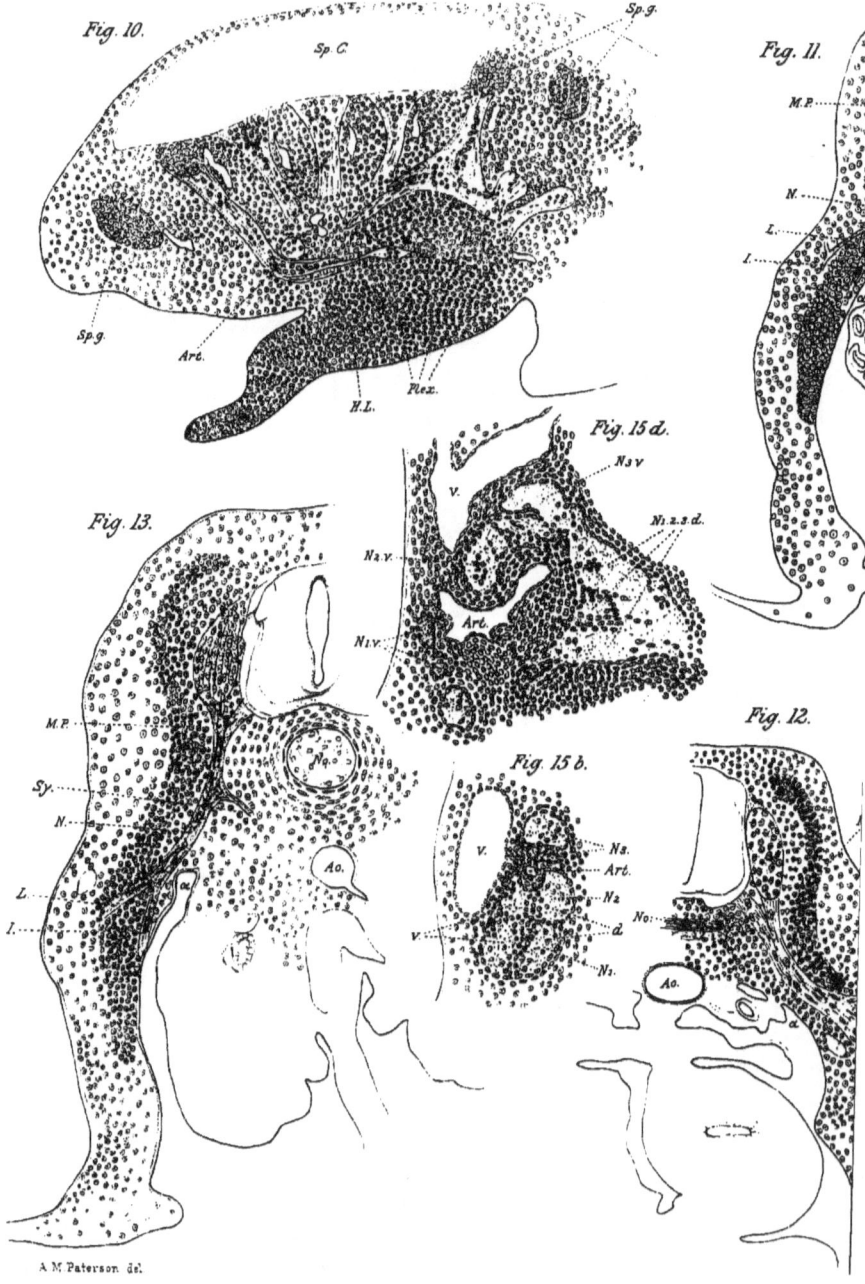

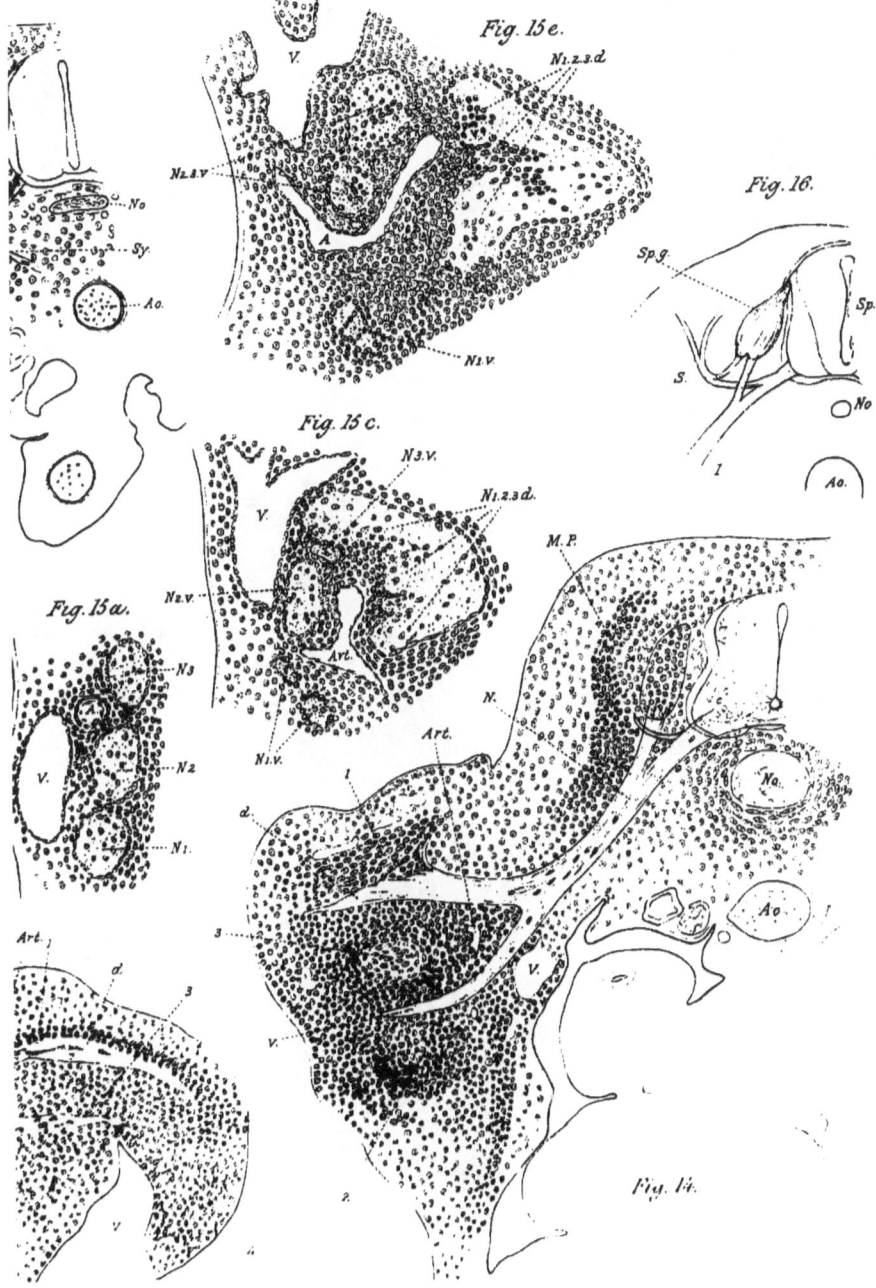

DEVELOPMENT OF THE FAT-BODIES IN RANA TEMPORARIA. A CONTRIBUTION TO THE HISTORY OF THE PRONEPHROS.

By ARTHUR E. GILES, B.Sc. (Lond.), M.B., Ch.B. (Vict.),

Platt Physiological Scholar, Owens College, Manchester; House Surgeon, Manchester Royal Infirmary.

[WITH PLATE IX.]

It has generally been held, since the researches of von Wittich, that the fat-bodies, or corpora adiposa, in the frog and allied Amphibians, are derived from the genital organs by a process of fatty degeneration in the anterior end of the primitive genital ridge.

Von Wittich himself says,* "they (the fat-bodies) have not at any time any connection with the Wolffian bodies, nor with the kidneys or their ducts." And again, "the genital organs become constricted into an anterior and a posterior part, of which the anterior becomes the fat-body, and the posterior the genital organ."

This investigation, conducted in the biological laboratories of Owens College, was begun with the intention of ascertaining if these views of von Wittich were correct. But its progress showed that the mode of development of the fat-bodies is very different from what von Wittich thought, and that the changes which take place are of a very interesting nature.

On dissecting a tailed frog, as represented in Fig. 1, the fat-bodies were seen already beginning to take on the lobose form which

---

* "Beitrage zur morphologischen und histologischen Entwickelung der Harn und Geschlechts Werkzeuge der nackten Amphibien," "Zeits. für wiss. Zool.," 4te Band, 1853, pp. 148, 149.

characterises them later on, and having in this case a peculiar resemblance to the fingers of a hand, as represented in the right half of Fig. 2. Thinking that the light-coloured bodies from which they sprang were the genital organs, these were removed with the surrounding parts, and cut horizontally in successive sections. It was then found that what had been taken macroscopically for genital organs (the real genital organs not appearing so plainly as in Fig. 1) showed microscopically most typical kidney structure, whilst at the same time it was quite continuous with the fat-body at the anterior end.

Two questions naturally arise: (1) How come the fat-bodies to be in relation with the anterior end of the kidneys? (2) How does the transition from this condition to that found in the adult take place? These questions I propose to answer in the following account.

Up to the age with which we are concerned, the generative cells are found in the condition of primordial ova, as described by Balfour;* hence there is no differentiation into ovary and testis. I shall therefore uniformly use the neutral term "genital organ."

The method adopted was to begin with very young tadpoles, and cut series of sections at various stages and in various planes, with the following results.

In a tadpole 8mm. long, that is, soon after the first appearance of the external gills, the three primitive openings of the pronephros into the body cavity can be seen. The tubules forming the pronephros are actually of larger diameter than they are somewhat later. One of them is represented in Fig. 3, in which it is seen that the cells lining the tubules are cubical or columnar, granular at the part nearest the lumen, and showing a distinct radial striation peripherally. The nucleus is central, and stains readily, as does also the nucleolus. The genital organ at this stage is situated nearer the median line than the pronephros, and anterior to it, and is well defined both anteriorly and posteriorly.

The young tadpole at the stage we are considering has still a plentiful supply of food-yolk, and is consequently independent of nutrition obtained from without. But now, as it grows, the absorption of the food-yolk proceeds more rapidly, while at the same

---

* "Comparative Embryology," vol. ii., p. 747.

time certain changes are observed, notably the gradual atrophy and subsequent disappearance of the external gills, while the recently acquired internal gills take on more work.

If at this stage the pronephros be examined, it will be seen that the tubules have a narrower diameter than those of the younger tadpole, while the cells are not so clearly defined. By this time the number of funnel-like openings into the body cavity has increased from three to five, and a new and important structure has made its appearance, the mesonephros, developed, as Sedgwick has shown,* in the mesoblast independently of the peritoneal epithelium.

The meso- and meta-nephros are not distinct from one another in the tadpole; they together form the kidney as found in the adult, and it is in this sense that the word kidney will be used.

The mesonephric tubules extend gradually from behind forwards till they come in contact with the pronephros. The whole nephros then acquires a distinct capsule, becomes separated from the muscular substance of the lateral mass, and lies freely in the abdominal cavity on the ventral aspect of the vertebral column, the peritoneum passing over it. Between the two kidneys is the aorta (Fig. 5). The genital organs, which arise as two hollow ridges, also gradually separate from the body wall, lying internal and ventral to the kidneys (Fig. 5), and are still perfectly well defined anteriorly, the proper genital substance extending quite to the anterior end.

Concurrently with these changes of conformation, the structure of the pronephros has been undergoing modification of the nature of a fatty degeneration. At the time that the hind limbs are just making their appearance, the degeneration has gone on to the extent represented in Fig. 4.

The way in which this conversion of kidney parenchyma into fat takes place is a true fatty degeneration, and not simply a fatty infiltration, though the latter occurs in the first stage. The change is seen best in the cells lining the glomeruli and renal tubules. The clearly defined margins of the cells become hazy, and the nuclei less distinct; fatty droplets appear at various parts of the cell and run together. The cells do not, however, swell up as the fatty matter

---

* "On the Early Development of the Anterior Part of the Wolffian Duct and Body in the Chick, together with Some Remarks on the Excretory System of the Vertebrata," "Quart. Journ. Micr. Sci.," vol. xxi., N.S., 1881, p. 449.

invades them, but their protoplasm becomes replaced by it. At a later stage the contents of the cell consist only of fatty granules and granular detritus.

For a while the outlines of the convoluted tubules can still be made out as in Fig. 6, the line of distinction between normal and degenerated kidney being well marked. But soon all trace of structure disappears, and there remains only a uniformly granular looking mass, as in Fig. 2. *This is the fat-body, or corpus adiposum.*

We have thus answered the first question that we proposed, "How come the fat-bodies to be in relation with the anterior end of the kidneys?" There further remains to be considered the question, "How does the transition from this condition to that found in the adult take place?"

When the hind limbs of the tadpole have appeared they develope fairly rapidly, the fore limbs sprouting out somewhat later. The condition of the "tailed frog" is now attained. While this is going on a change takes place in the urino-genital organs, which, as regards the time at which it occurs, varies somewhat in different tadpoles, but usually begins during the period in which the tail is commencing to atrophy, and is for the most part completed by the time the tail is quite absorbed.

This change is as follows: The anterior end of the nephros grows ventrally, and becomes secondarily attached to the anterior end of the genital organ, ovary or testis, as the case may be. Fig. 7 shows the urino-genital organs during this intermediate stage of transition, the fat-body (f) being directly continuous both with the kidney (k) and with the genital organ (g).

This occurs at about the time that the mesonephric tubules are growing out towards the genital organ, forming the future vasa efferentia in the case of the male. Thus, in some sections the condition of this double outgrowth from the excretory to the genital organs can be seen, the three parts of the nephros, pro-, meso-, and meta-, being quite continuous.

Ultimately the attachment of the fat-body to the kidney gives way, and the former remains attached to the anterior end of the genital organ, as it is in the adult (Figs. 8 and 9). We thus see that the fat-body is not the anterior end of the genital organ, which has undergone fatty degeneration, as was thought by von Wittich, but that its attachment to the genital organ is secondary.

The fatty degeneration is always complete before the attachment to the genital organ takes place; almost any tailed frog that has not very long had its four limbs showing the fat-body attached to the anterior end of the kidney. That it is, in reality, fat-body, is shown by its macroscopic and microscopic characters, and by its staining with osmic acid. The part marked (f) in Fig. 7 has exactly the same structure and appearance as that similarly marked in Fig. 8, the two specimens having been stained and cut at the same time.

Having thus decided that the fat-bodies are derived not from the genital organs but from excretory structures, we have to consider what part of the nephros it is to which they owe their origin. It can only be pro-, meso-, or meta-nephros, or their ducts; the ducts can be at once put aside, because their destination has been clearly and definitely made out. The meso- and meta-nephros are also known to form together the permanent kidney, as found in the adult.

There remains, therefore, only the pronephros, which, in the Amphibians at least, has hitherto received but little attention, though Sedgwick* mentions that it undergoes atrophy in the young frog. "Atrophy," however, implies diminution in size, or even total disappearance; the pronephros of the tadpole, on the contrary, not only persists but actually gets larger (in its modified form) as the frog grows. Now, we saw that at an early stage the pronephros undergoes a fatty degeneration; that the degenerated part remains for a time continuous with the rest of the kidney (Fig. 6), and then becomes secondarily attached to the genital organ. Hence *the fat-bodies represent the persistent pronephros, profoundly modified both in structure and in function.*

If it be objected that it is *a priori* improbable that the fat-bodies should consist of the anterior part of the nephros detached and fastened on secondarily to the genital organ, it will be sufficient to recall the fact that the vasa efferentia are formed by a quite parallel growing out of kidney structure—the mesonephric tubules; the only difference being that the process in the case of the fat-

---

* Op. cit., p. 445.

body goes a step farther, since the primary connection with the kidney is lost, while the vasa efferentia remain connected with both kidney and testis.

Again, the question may be asked, "Why should only a part of the kidney structure undergo fatty degeneration—why should not the meso- and meta-nephros share in the change ?" The answer would be even more difficult to find if, on the supposition that von Wittich was right, such a question were asked concerning the genital organs, for they are of equal value in all their parts, and when the metamorphosis occurs no portion of them has had any reproductive activity. But in the case of the pronephros it is different. It is true that the nature and origin of the pronephros are still matters of discussion, but it is at least evident that the pronephros is in many respects different from the mesonephros ; that the former, in the case of the frog and of all animals with a larval stage, has a period of activity before the mesonephros appears at all, and in most cases disappears as the latter begins to take on active functions. On the other hand, in Vertebrates possessing no larval stage, the existence of the pronephros is only dimly shadowed forth by rudimentary traces, the meso- and meta-nephros performing all the excretory functions from the first.

The answer then to the question, "Why this change should occur resulting in the formation of the fat-body," seems to be this—that with the close of larval life the pronephros is no longer needed, and in harmony with the pathological law that atrophy follows disuse, it degenerates to the condition of fat-body. Doubtless, however, this law is here so far modified that the fat-body still serves some useful purpose in the organism, though what that purpose is is not at all clear. It is in all probability an example of "change of function," the later function being in some way nutritive.

As to the distribution of fat-bodies—they are unknown outside the Amphibian group. According to Stannius, Hoffmann, Wiedersheim, and others, they are present in all Amphibians. We have very little knowledge of their function beyond that they are concerned in all probability with nutrition, serving as a reserve stock at certain times of the year. They are differently placed in the several groups in which they occur, and it is by no means certain whether they are homologous structures in all cases.

The fate of the pronephros in the frog, as above described, throws some light on the condition that obtains in other groups of Vertebrates.

It was stated by Balfour* that "the pronephros atrophies more or less completely in most types, though it probably persists for life in the Teleostei and Ganoids."

In a later paper,† however, after working over the condition of the kidneys in the sturgeon and in certain Teleostei, he stated that "the whole of the apparent kidney in front of the ureter, including the whole of the so-called head kidney, is simply a great mass of lymphatic tissue, and does not contain a single uriniferous tubule or Malpighian body," from which he concluded that both in Ganoids and in Teleostei the organ usually held to be pronephros is actually nothing of the kind. He therefore considered that Rosenberg‡ was mistaken in thinking that he had traced in the pike the larval organ into the adult part of the kidney called by Hyrtl the pronephros; and his final conclusion was "that the pronephros, though found in the larvæ or embryos of almost all the Ichthyopsida, except the Elasmobranchii, is always a purely larval organ, which never constitutes an active part of the excretory system in the adult state." But Balfour did not apparently regard it as possible that the pronephros might continue in the Ichthyopsida in a modified condition, but thought that if it did not persist with at least its original structure, if not its original function, it must have disappeared altogether. He was, however, led to this conclusion by the study not of their development, but of their adult structure.

But it seems to me, from a consideration of the state of things in the tadpole and young frog as above described, that it is not at all necessary that the pronephros, if it persist, should retain its original structure any more than its original function; that it is quite possible that Rosenberg's observations were correct, since the only argument adduced against them is this alteration of structure, and that there is nothing in Balfour's observations on the Ganoids and

---

* " Comparative Embryology," vol. ii., p. 729.

† " On the Nature of the Organ in Adult Teleosteans and Ganoids which is usually regarded as the Pronephros or Head Kidney," "Quart. Journ. Micr. Sci.," vol. xxii., N.S., 1882.

‡ " Untersuchungen über die Entwicklung der Teleostierniere," Dorpat, 1867.

K

Teleosteans to contradict them. The fate of the pronephros in Teleosteans and Ganoids is, from this standpoint, closely analogous to that in the tadpole, except that in the latter it undergoes yet further modification in becoming quite separated from the true kidney and attached permanently to the genital organ.

The fact that the pronephros does persist in a modified form seems to me in nowise to detract from but rather to add to the probability of Gegenbauer's views being correct, namely, that the pronephros is the primitive excretory organ of the Chordata, and that its substitute in existing Vertebrata, the mesonephros, is phylogenetically a more recent organ.

I may sum up my conclusions as follows :—

1. The fat-bodies in the frog, and hence presumably in allied Amphibians, are formed by a fatty degeneration, not of the anterior end of the genital organs, but of original kidney structure.

2. The part of the kidney which undergoes this conversion into fat-body is the pronephros or head kidney.

3. It seems very probable from analogy, and from the researches of Rosenberg, that the structure in front of the true kidney in Ganoids and Teleostei, described by Balfour as lymphatic tissue, is the persistent but structurally and functionally modified pronephros.

4. The fact that a part of the kidney undergoes such a remarkable change, the rest remaining normal and functional, is an additional argument in support of the view that the pronephros has a different phylogenetic history from the mesonephros, and that it is more ancestral.

It only remains for me to perform the pleasant duty of expressing my warm thanks to Professor A. Milnes Marshall for the uniform and stimulating kindness with which he has helped me in this short research by suggestions and criticisms; he has been good enough to go over my specimens with me, and to discuss with me my results.

I desire also to express my obligations to my friend Dr. G. Herbert Fowler for much valuable and practical assistance.

My thanks are further due to Professor Stirling, under whose direction the work has been done.

DESCRIPTION OF PLATE IX.

Illustrating Mr. A. E. Giles's Paper on "The Development of the Fat-Bodies in Rana Temporaria."

The letters have the same significance in all the figures. *b.* Muscles of body wall on the ventral aspect of the vertebral column. *f.* Fat-body. *g.* Genital organ (sex undifferentiated). *k.* Kidney. *n.* Notochord.

Fig. 1. Tailed frog, dissected so as to expose the urinogenital organs. The kidneys are seen lying against the vertebral column, and continuous anteriorly with the fat-bodies. Anterior and internal to the kidneys are the genital organs ($\times$ 4).

Fig. 2. Anterior end of the urinogenital organs of the tailed frog shown in Fig. 1, enlarged. The right half of the figure shows the surface view, the left half shows the appearance in horizontal section ($\times$ 50).

Fig. 3. Normal pronephric tubule, from a tadpole still possessing external gills ($\times$ 350).

Fig. 4. Pronephric tubules showing fatty degeneration, from a tadpole whose hind limbs were just appearing ($\times$ 350).

Fig. 5. Transverse section through the lumbar region of a tailed frog, showing the mode of development of the genital organs and their relation to the excretory organs at this stage ($\times$ 50).

Fig. 6. Sagittal section through the lumbar region of a tadpole that had recently acquired its fore limbs, showing the anterior end of the nephros partly degenerated ($\times$ 60).

Fig. 7. Sagittal section through the lumbar region of a tailed frog whose tail had begun to be absorbed, showing the fat-body connected with both kidney and genital organ ($\times$ 60).

Fig. 8. Sagittal section through the lumbar region of a young frog that had just lost its tail ($\times$ 60).

Fig. 9. A young frog at the same stage as the preceding, dissected so as to expose the urinogenital organs, which present the same condition as in the adult ($\times$ 4).

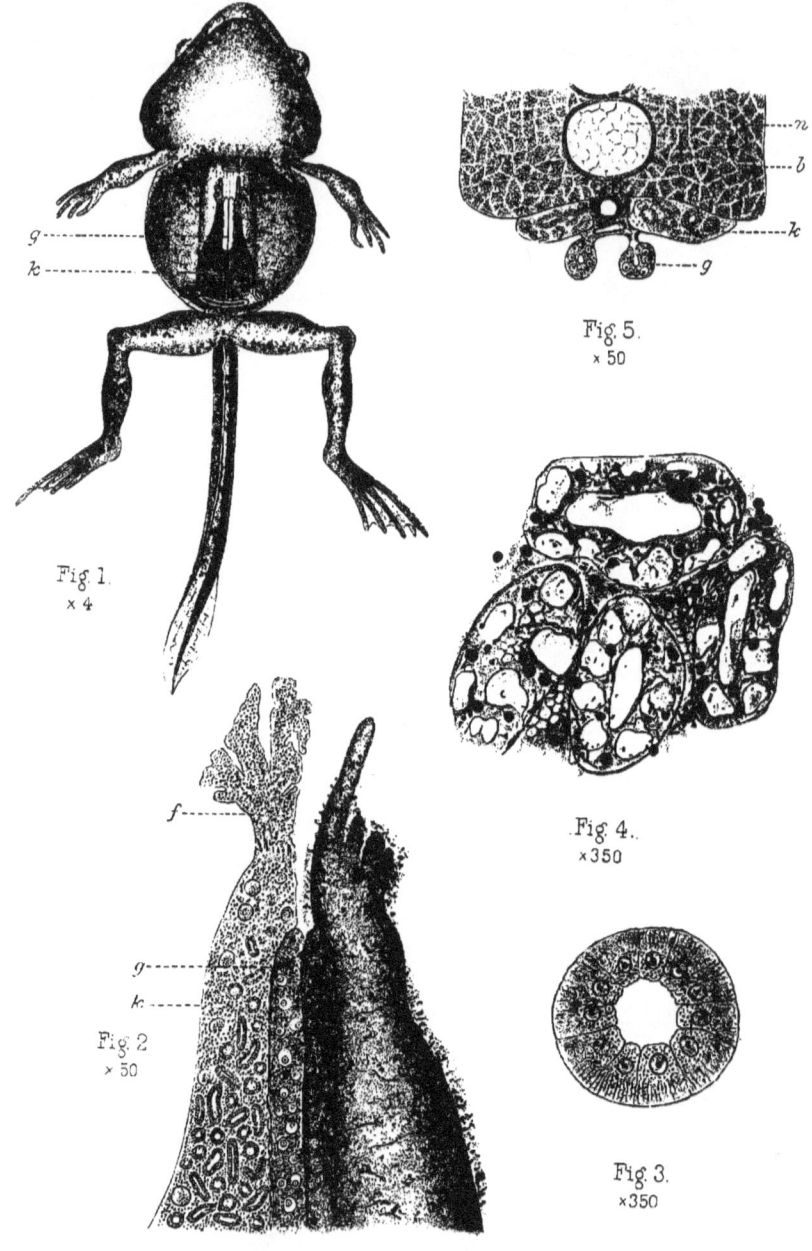

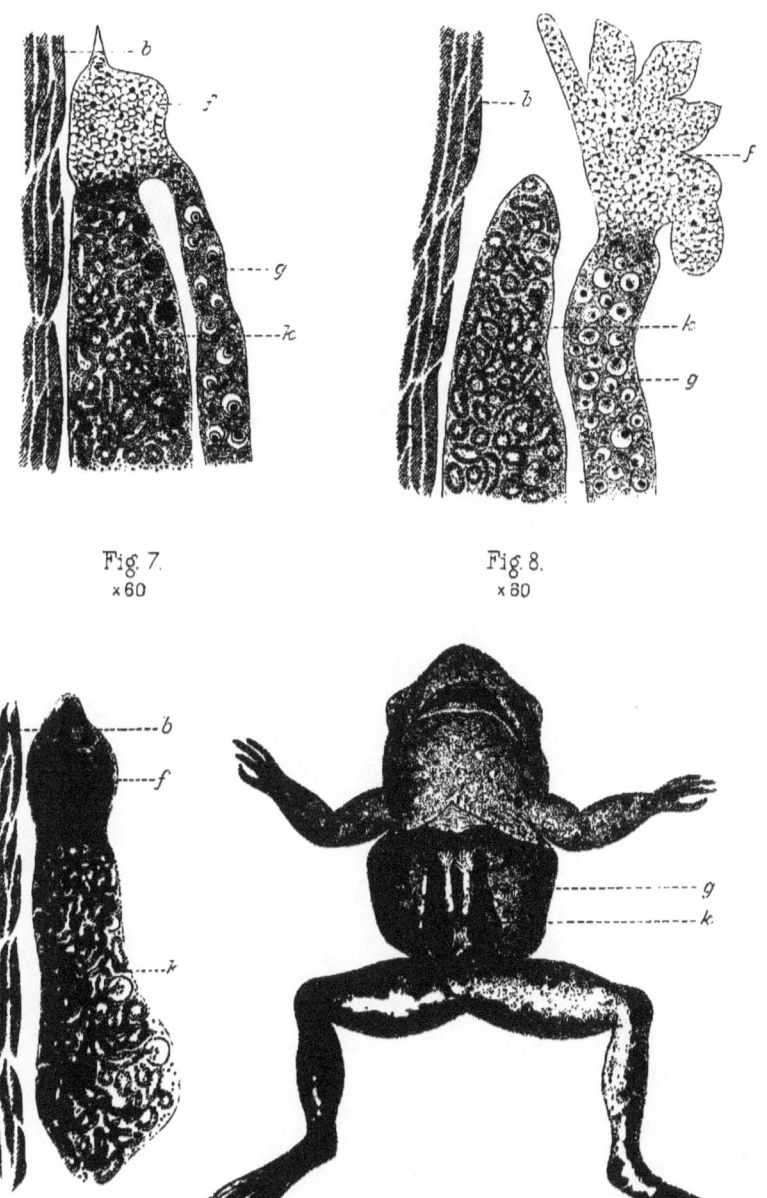

Fig. 7.
×60

Fig. 8.
×80

Fig. 8.
×80

Fig. 9.
×4

THE DEVELOPMENT OF THE KIDNEYS AND FAT-BODIES IN THE FROG.

By A. MILNES MARSHALL, M.D., F.R.S., *Beyer Professor of Zoology in the Owens College;* and
EDWARD J. BLES, *of the Owens College.*

[WITH PLATE X.]

IN the present paper we do not propose to attempt a complete description of the development of the urinary organs in the frog, but shall confine ourselves to certain points which have recently engaged our attention. Of these, the more important are : (1) the mode of formation of the head kidney and its duct ; the structure and relations of these parts during the successive stages of larval life ; and the degenerative changes which they undergo about the time of the metamorphosis : (2) the development of the tubules of the Wolffian body : (3) the development of the nephrostomes and their anatomical relations during larval life and in the adult condition : (4) the early development of the fat-body. All our observations were made on specimens of *Rana temporaria*, and are based partly on dissections, partly on the examination of sections in the three principal planes, transverse, sagittal, and horizontal.

I. THE HEAD KIDNEY AND ITS DUCT.

1. *Historical Account.*—Goette[*] gives a careful account of the early stages in the development of the head kidney and its duct in Bombinator. He describes the head kidney as arising in the first instance as an outward fold of the somatopleuric mesoblast at the

---

[*] Goette, "Die Entwickelungsgeschichte der Unke," 1875, pages 819-828, and plates vi., vii., xiii., and xxii.

anterior end of the trunk, immediately ventral to the muscle-plates. This fold is at first ill defined dorsally, but sharply marked along its lower or ventral edge. At a slightly later stage the fold, which lies immediately beneath the epiblast, becomes more clearly marked off from the rest of the mesoblast, and its communication with the cœlom or body cavity becomes reduced to a narrow longitudinal slit confined to its anterior end, extending along about three myotomes. The hinder part of the fold forms a tube which extends backwards towards the cloaca, lying between the epiblast and somatopleuric mesoblast. By further changes, the slit-like aperture of communication with the cœlom becomes divided by constrictions into three tubular openings, which become the three nephrostomial apertures characteristic of the Anuran head kidney.

The longitudinal tube into which the three nephrostomes open, become twisted on itself like a letter S, and become further complicated by the development of lateral diverticula from the limbs of the S; the complicated and convoluted tubular mass so formed constituting the head kidney. The hinder part of the tube continues its growth backwards, and soon acquires an opening into the cloaca.

Goette notes further that the cœlom first becomes a conspicuous cavity opposite to the head kidney, and that into this cavity the glomerulus projects on each side as a thick walled lateral diverticulum from the aorta. Later on the cœlom extends ventralwards round the sides of the body by separation of the somatopleuric and splanchnopleuric layers of the mesoblast from each other. At a still later stage, when the lungs have attained some size, the part of the cœlom opposite the glomerulus and the nephrostomes becomes again marked off, though not completely separated from the rest of the body cavity, by fusion of the outer surface of the lung with the somatopleure at the level of the head kidney.

Fürbringer,* in his admirable account of the development of the Amphibian kidney, gives a useful summary of the work done by previous investigators, and a careful description of the mode of development of the head kidney and its duct in *Rana temporaria*, as well as in *Salamandra* and other Amphibians. He agrees in the main with Goette, but corrects him in certain points, and adds a number of

---

* Fürbringer, "Zur Entwickelung der Amphibienniere," Heideberg, 1877, pp. 13-32.

further details. He describes the head kidney and its duct as arising as a groove in the somatopleure, the lips of which close to form a tube; this tube opens in front into the cœlom by a longitudinal slit-like mouth, which by fusion of its lips at two places gives rise to the three nephrostomial apertures of the head kidney. Of these three apertures, the two anterior ones lie close together, the first at a more ventral level than the second; the third aperture is some little distance behind the other two.

Owing to unequal rates of growth of the various parts, this anterior portion of the tube, with the three nephrostomes, becomes situated at a more dorsal level than the hinder part of the tube, which now opens into it, not as at first behind the third nephrostome, but on its ventral surface between the second and third nephrostomes. This hinder portion of the duct rapidly elongates and forms an S-shaped loop lying ventral to the nephrostomes. In the later stages the convolutions become more marked, and are further complicated by lateral diverticula, of which there are three principal ones, which intertwine amongst the convolutions of the main tube and give off secondary branches, which, like the main diverticula, end blindly. In this way the head kidney is formed as a complexly twisted tubular gland, roughly spherical in shape, and imbedded in the somatopleure at the anterior end of the body cavity. The nephrostomial apertures become drawn out into tubes with the growth of the gland, but retain their openings into the cœlom so long as the gland itself exists. The ducts soon acquire openings into the cloaca at their posterior ends; along the greater part of their course they are flattened dorso-ventrally, and irregularly swollen at intervals. Both the head kidney and its duct are, according to Fürbringer, distinctly and exclusively mesoblastic structures.

Selenka* gives an account of the structure and relations of the head kidney and its duct in two young embryos of Hylodes Martinicensis. He had no opportunity of working out the details of development; but his figures, reconstructed from a series of sections, are of great excellence, and by far the most instructive that have yet been published. The extreme irregularity in size of the tubules of the head kidney, and the marked asymmetry between the organs

---

* Selenka, "Der Embryonale Secretionsapparat des Kiemenlos Hylodes Martinicensis," Berlin, 1882.

of the two sides, are conspicuous features in his figures, and will be referred to again further on.

Duval's* description of the development of the head kidney and its duct in the frog is illustrated by a series of useful figures, but contains no new matter of importance. In all essential points he agrees with Fürbringer.

Wichmann† also confirms Fürbringer's account.

Hoffmann‡ adds a few details to the previous description. He notes that the cells lining the nephrostomial tubes bear flagella, and not cilia as sometimes stated. Each cell bears a single flagellum, which is considerably longer than the diameter of the tube in which it lies, and is normally directed along the axis of the tube, with its free end pointing inwards, *i.e.*, towards the kidney and away from the peritoneal opening of the tube.

The most recent paper on the development of the Amphibian head kidney with which we are acquainted is a short note by Mr. J. H. Kellogg,§ in which the following passage occurs: "The most important point to be noticed is the formation of the so-called ventral part of the gland. It has always been described, at least in Amphibia, as being formed from that part of the duct immediately behind the last or most posterior nephrostome. . . . In Amblystoma, and also in the frog, this lower portion is formed from the ventral side of the dorsal part of the pronephros and anterior to the last nephrostome." Mr. Kellogg's description is perfectly correct, but it is also in exact accordance with the accounts given by Fürbringer and Goette, the recognised authorities on the subject, and not, as he supposes, at variance with these.

2. *The Early Development of the Head Kidney and its Duct.*—We have worked over these early stages carefully; but inasmuch as our results agree in almost all points with Fürbringer's, we do not propose to describe them in detail.

---

* Duval, "Sur le Développement de l'Appareil Génito-Urinaire de la Grenouille. 1re Partie—Le Rein Précurseur," Montpellier, 1882.

† Wichmann, "Beiträge zur Kenntniss des Baues und der Entwicklung der Nierenorgane der Batrachier," Bonn, 1884.

‡ Hoffmann, "Zur Entwickelungsgeschichte der Urogenitalorgane bei den Anamnia," "Zeitschrift für wissenschaftliche Zoologie," Bd. 44, 1886.

§ Kellogg, "Notes on the Pronephros of Amblystoma Punctatum," "Johns Hopkins University Circulars," vol. ix., No. 80, Baltimore, April, 1890, p. 59.

We find that in tadpoles of 3½ mm. to 4 mm. length the head kidney is just commencing to form, and is present as a longitudinal fold of the somatopleure at the anterior end of the body, and immediately below the ventral borders of the myotomes. The fold extends back some distance on either side as the rudiment of the duct, but does not as yet reach the cloaca.

As regards the formation of the nephrostomes, the relations between the nephrostomes and the convoluted part of the duct, the outgrowths of blind diverticula in the head kidney, and the growth backwards of the duct towards the cloaca, our observations agree exactly with Fürbringer's.

We find that the duct developes *in situ* along its whole length, and not by growth backwards from its anterior portion. We also find that though it lies very close to the surface epiblast, yet that the epiblast takes no share whatever in its formation.

At the time of first appearance of the head kidney, the body cavity or cœlom is a potential rather than an actual space between the parietal and splanchnic layers of the mesoblast. By the time the three nephrostomes are established, the dorsal part of the body cavity, opposite to these and immediately below the myotomes, becomes dilated, and into it projects on either side the glomerulus of the head kidney, a thick walled sacculated diverticulum from the dorsal aorta. This dorsal portion of the body cavity, with the glomerulus projecting into it from the inner or median side, and the nephrostomes leading outwards from it into the tubules of the head kidney, has been compared to the Bowman's capsule of a Malpighian body, a comparison based at present on very insufficient evidence. Some stress has been laid on the fact that at this early period, about the time of hatching of the tadpole, this dorsal portion of the body cavity does not communicate freely with the ventral part of the body cavity; but this in the frog is simply due to the mass of the food-yolk, which distends the splanchnopleure and keeps it in close contact with the somatopleure, preventing the formation of a conspicuous body cavity in this region until a considerable portion of the yolk has been absorbed.

The segmental arrangement of the three nephrostomes of the frog's head kidney has been a subject of discussion. We find that a true segmental arrangement is present in the early stages. The first or most anterior myotome in the frog extends forwards to the level of the hinder border of the auditory vesicle in a tadpole of about $4\frac{1}{2}$ mm. length. The first nephrostome lies immediately below the ventral border of the second myotome, close to its posterior edge; the second nephrostome is rather more dorsally placed, and lies below the ventral border of the third myotome, rather nearer its posterior than its anterior edge; and the third nephrostome is similarly related to the fourth myotome.

The nephrostomes are at first mere holes leading from the body cavity into the dorsal part of the tubular head kidney. With growth of the head kidney, these holes become drawn out into tubes, which may conveniently be spoken of as the nephrostomial tubes or tubules (Fig. 6, N T). These nephrostomial tubules differ in many respects from the ordinary tubules of which the rest of the head kidney consists; they are of much smaller calibre, and their walls consist of a very thin outer connective tissue coat, within which is a single layer of cubical epithelial cells, which have a curiously clear, transparent appearance, have strongly marked and pigmented walls, and bear long flagella, which project into the tubule, with their free ends directed away from the body cavity.

The duct of the head kidney acquires its opening into the cloaca in tadpoles between $4\frac{1}{2}$ mm. and 5 mm. in length, *i.e.*, just previous to the time of hatching. The terminal portion of the duct appears to us to be developed rather as a lateral outgrowth from the cloaca than by further growth on the part of the duct itself.

Concerning the nomenclature of this duct, it is exceedingly difficult to be consistent. At the time the tadpole hatches, the head kidney is a well-developed convoluted tubular gland, and is the only renal organ present; the duct is clearly connected with it, and with it alone; and the closeness of the connection is emphasized by the continuity of the two structures from the time of their first appearance. It seems, therefore, reasonable to speak of the duct as the duct of the head kidney, or pronephric duct. Difficulties, however, arise in the later stages, when the hinder part of the duct receives the tubules of the Wolffian body; while in the adult frog the head kidney disappears completely, and the posterior part of the duct persists in both sexes as the ureter. Duct of the head kidney, or pronephric duct, become clearly unsuitable terms in these later stages, and either segmental duct or archinephric duct becomes preferable.

3. *The Changes undergone by the Head Kidney and its Duct.*—The head kidney is the excretory organ of the tadpole during the earlier part of its existence. At the time of hatching, the head kidney is already convoluted, and has acquired its three nephrostomial openings into the body cavity, while its duct opens behind into the cloaca, and through the cloacal aperture to the exterior.

The mouth is not formed till a much later period, and inasmuch as until the mouth perforation is completed, no food can be taken into the alimentary canal, it seems not improbable that the early formation of the cloacal opening is for the sake of providing an exit for the excretory matters separated by the head kidneys. The large size of the head kidneys and their further rapid growth certainly point to their being in physiological activity from the time of hatching of the tadpole, and the histological characters of the epithelial cells of the tubules afford corroborative evidence.

After hatching, the head kidneys for a time grow rapidly, and become extremely complicated, as described in the last section. They receive an abundant vascular supply from the posterior cardinal veins, which surround them, and send in branches between the individual tubules. The head kidneys attain their greatest size in tadpoles of about 12 mm. length, in which the opercular folds have grown back over the gills, and the hind limbs are just visible as little buds at the sides of the anal spout. At this stage the tubules of the Wolffian body are beginning to develope: as these increase in number and in size they gradually replace the head kidneys functionally, while these latter undergo degenerative changes and ultimately disappear completely.

In tadpoles of from 12 to 18 or 20 mm. in length, the head kidneys remain of about the same size, and show but slight structural changes; but in tadpoles of from 20 mm. to 24 mm. in length, by which time the Wolffian bodies have attained considerable size and complexity, the head kidneys, though as large as before, or nearly so, have undergone important degenerative changes.

Transverse sections at the latter period (Fig. 6) show that the tubules of the head kidney are enormously dilated at places (Fig. 6, N C), forming irregular cavities as large in transverse section as the lungs;

in other parts the tubules are of very irregular size, and often laterally compressed. The walls of the tubules, and more particularly of those which are most dilated, have also undergone important changes. While in the earlier stages each tubule is lined by a single layer of cubical cells, with clearly defined outlines, at the present stage the epithelial cells are cloudy in appearance, with ill-defined and often indistinguishable boundaries, while their inner surfaces, facing towards the cavity of the tubule, have exceedingly ragged and irregular outlines. Within the tubules are found at intervals larger or smaller masses of débris, apparently resulting from the breaking down of some of the epithelial cells. The three nephrostomes are still present (Fig. 6, N T), and communicate with the tubules by narrow nephrostomial tubes lined by pigmented epithelial cells bearing flagella, as in the earlier stages.

The dilatation of the tubules is most marked, not in the main tubes, but in the lateral diverticula, which, we have seen, end blindly. It is worthy of notice, that although the epithelium of the tubules is clearly degenerating, yet that there is no sign whatever of fatty degeneration, such as has been described by Duval[*] and other writers. The appearances rather suggest that the changes are caused by accumulation of fluid within the tubules, and especially their blind diverticula, due perhaps to obstruction, partial or complete, of the duct; the result of this accumulation of fluid being first dilatation of the tubules, and then disintegration of their living epithelium.

A comparison of our specimens and figures with the figures given by Selenka[†] of the head kidney in the tadpole of Hylodes Martinicensis at a slightly earlier stage of development, strongly suggests that in this latter case also the tubules are in the state of dilatation immediately preceding degeneration and atrophy. Selenka figures the tubules as of very unequal size, the lateral diverticula being especially large.

From this stage the head kidneys of the tadpole steadily degenerate. In tadpoles of 40 mm. length they are still spherical bodies of fair size (Fig. 1, N A), though much smaller both relatively and absolutely than before. All three nephrostomes are present, and open freely into the tubules. The tubules are irregular

---

[*] Duval, loc. cit., p. 28.
[†] Selenka, loc. cit., Figs. 1, 2.

in size, but show no marked dilatations such as are present at the earlier stage.

During the metamorphosis the degeneration of the head kidney proceeds rapidly; the whole organ shrinks greatly in size, the cells lose their outlines, and pigment appears in and between them. As was first shown by Hoffmann,* the first, and a little later the second, nephrostome closes up and disappears; the third, or most posterior nephrostome, persists for a time, but ultimately shares the same fate and disappears.

The condition of the head kidney in a frog shortly before the disappearance of the tail is shown in Fig. 5, N A. At this stage the head kidney consists of a small group of narrow convoluted tubules with pigmented walls; the two anterior nephrostomes have disappeared, but the third or hindmost one is still present, and opens into the body cavity.

In a frog at the end of the first year the head kidney is almost obliterated, and consists merely of a few indistinct cells surrounded by pigment: all three nephrostomes have disappeared.

Concerning the changes undergone by the archinephric duct, or duct of the head kidney, our own observations are very imperfect, but are perhaps worth recording, inasmuch as they confirm, so far as they go, the only account with which we are acquainted of the development of the Wolffian and Müllerian ducts in the frog, *i.e.*, the exceedingly careful and valuable description given by Hoffmann in the paper already quoted.†

The archinephric duct is, as we have seen, from the first in direct connection with the head kidney. On the formation of the tubules of the Wolffian body, which first appear in tadpoles of about 12 mm. length, these tubules open into the duct, so that the archinephric duct for a time acts as the duct for both the head kidney and the Wolffian body. At the time when the head kidney begins to degenerate (Fig. 6), the part of the archinephric duct between the head kidney and the anterior end of the Wolffian body (*cf.* Figs. 1 and 8, N D) becomes much flattened dorso-ventrally, and its lumen almost completely obliterated; and it is very possible that, as suggested above, this obstruction or obliteration of the lumen of

---

* Hoffmann, loc. cit., p. 594.
† Hoffmann, loc. cit., pp. 594–599.

the duct is an important factor in causing the dilatation of the tubules of the head kidney, which forms the first stage in their degeneration.

The succeeding stages we have only followed imperfectly. According to Hoffmann, who lays stress on the great difficulty of the investigation, they occur as follows:—Towards the close of the metamorphosis, at a period when the greater part of the tail is absorbed, *i.e.*, a stage corresponding to our Fig. 5, the archinephric duct separates completely from the head kidney, which latter still communicates with the body cavity through the third or hindmost nephrostome. The peritoneal epithelium opposite the degenerating head kidney becomes columnar, the patch of columnar cells extending outwards as a narrow transverse strip some distance beyond the outer edge of the head kidney. With this columnar epithelium the blind anterior end of the archinephric duct, which is now completely separate from the head kidney, comes in contact, and then fuses.

After the completion of the metamorphosis and the entire disappearance of the tail, the part of the archinephric duct in front of the Wolffian body splits into two, which become respectively the Wolffian duct and the Müllerian duct. Of these, the Wolffian duct commences blindly in front, and runs backwards through the Wolffian body as the ureter, receiving the collecting tubules of the kidney, and in the male the seminal tubules as well. The Müllerian duct is at first short, and lies entirely in front of the Wolffian body or kidney; its anterior end is blind, and fused as described above with the columnar peritoneal epithelium opposite to the head kidney; its posterior end is also blind, and lies to the outer side of the anterior end of the Wolffian duct.

The anterior end of the Müllerian duct soon acquires an opening into the body cavity. This opening lies close to, but is independent of, the third nephrostome, which is still present, but which closes and disappears very shortly afterwards. The anterior opening of the Müllerian duct does not, as might be expected, become directly the mouth of the oviduct, but undergoes first a series of complicated changes. From the opening a groove extends outwards along the columnar strip of peritoneum already noticed; by closure of the lips of this groove it becomes converted into a tube continuous at its inner end with the Müllerian duct. Later on, through unequal growth of the surrounding parts, this tube shifts its relations, and in place of running directly outwards, runs forwards, outwards, and downwards round the anterior border of the root of the lung, ending with a backwardly directed mouth, which opens into the body cavity below the lung. Still later, the greater part of the length of this tube opens out by separation of its lips, and becomes once more a groove, which gradually flattens and ultimately disappears altogether, while its dorsal end opposite the root of the lung becomes the permanent mouth of the oviduct.

The hinder end of the Müllerian duct grows backwards, according to Hoffmann, quite independently of the Wolffian duct, to the outer side of which it lies, but in very close connection with a longitudinal strip of peritoneum, the cells of which are columnar in shape. The actual mode of growth of the duct was not observed, but Hoffmann is inclined to regard is as effected at the expense of the strip of peritoneal cells.

Of the above account by Hoffmann we can confirm certain portions.

In a tailed frog, such as that shown in Fig. 5, the part of the archinephric duct between the head kidney and the Wolffian body is extremely slender, and at places its lumen is entirely obliterated. We have, however, not seen any division of this duct into Wolffian and Müllerian ducts at this stage. Further back, on entering the Wolffian body, the duct becomes much larger, and has a conspicuous lumen.

In a frog at the end of the first year we find the head kidney almost obliterated and completely separate from the duct; all three nephrostomes have disappeared completely. The Müllerian duct is well developed at its anterior end, which runs, exactly as Hoffmann describes it, almost vertically downwards round the outer border of the root of the lung, and then backwards for a short distance along its ventral surface, ending with an open mouth into the body cavity, placed a short distance below the level of the pulmonary vein. We can, therefore, confirm Hoffmann's statements that in the frog the abdominal opening of the Müllerian duct is not formed from one of the nephrostomes, and that in a young frog the Müllerian duct at first extends round the root of the lung and opens on its ventral surface.

Concerning the posterior end of the Müllerian duct, we find that in frogs of this age, the end of the first year, the duct may be traced back as far as the anterior end of the Wolffian body. It lies to the outer side of the aorta, and is exceedingly slender along the greater part of its length; towards its hinder end the lumen disappears, and the duct becomes a solid rod of cells. We have traced this into the Wolffian body, but have failed to determine how it ends, or in what manner its further growth backwards is effected. We note, however, that it does not in our specimens lie quite so close to the peritoneum as Hoffmann describes and figures.

In a second year's female frog the Müllerian duct is formed along its whole length, but still has an exceedingly small lumen, or is actually solid along part of its course. In front, its opening into the body cavity is now in the position which it occupies in the adult.

Before leaving the head kidney, we may refer briefly to the glomerulus and the changes which it undergoes. The glomerulus of the head kidney arises as a sacculated outgrowth from the ventral and outer wall of the aorta, which bulges outwards into the body cavity opposite the nephrostomes. At the time of hatching of the tadpole, this dorsal part of the body cavity is, as already noticed, practically the only part present, for though the splitting of the mesoblast extends down the sides of the body to the ventral surface, yet the two layers, somatic and splanchnic, are in close contact, owing to the great mass of the food-yolk, except at this upper or dorsal angle. The glomerulus therefore appears at this period to be in a special cavity, which, later on, opens into the general coelom on absorption of the yolk and further development of the abdominal viscera. At a later stage still, as shown for a 23 mm. tadpole in Fig. 6, **T**, the part of the body cavity in which the glomerulus lies becomes partially boxed in by fusion of the outer wall of the lung with the peritoneal covering of the head kidney. The partition so formed is only an incomplete one, inasmuch as the part of the coelom lodging the glomerulus communicates freely with the general body cavity posteriorly, and also anteriorly in front of the root of the lung.

The development of the glomerulus keeps pace with that of the head kidney. It is large up to about 23 mm. (Fig. 6, **N G**), but from that stage commences to dwindle. Its size and relations at

40 mm. are shown in Fig. 1, N G, and in the tailed frog in Fig. 5, N G. It is still present, though very small, at the end of the first year, but is absent in frogs of the second year.

Its position directly opposite the nephrostomes of the head kidney, and the fact that its development, both progressive and retrogressive, keeps pace with that of the head kidney, points to a close physiological connection between the two organs; though it is not easy to imagine what precise function the glomerulus subserves.

II. THE DEVELOPMENT OF THE WOLFFIAN BODY OR KIDNEY.

1. *The Formation of the Wolffian Tubules.*—Like all previous observers, we have found the early stages in the development of the Wolffian tubules in the frog extremely difficult to determine, and our observations are by no means so complete as we could wish.

Fürbringer* describes the Wolffian tubules as arising in Urodeles in the form of solid ingrowths of the peritoneal epithelium close to the root of the mesentery. These form solid strings of cells, which soon separate from the peritoneum, acquire a lumen, and become convoluted tubes.

Hoffmann† describes the development in Urodeles somewhat differently. He agrees with Fürbringer in describing the earliest stage in the development of a Wolffian tubule as a solid ingrowth of peritoneal cells, but he differs from him in maintaining that the peritoneal connection is retained and gives rise to the nephrostome through which the Wolffian tubule opens into the cœlom. In this respect he is in agreement with Goette and with Spengel. In Anura, Hoffmann describes the early stages as identical with those of Urodeles, the nephrostomes arising as solid ingrowths of peritoneum, of which the inner ends become tubular, acquire openings into the archinephric duct, and then growing rapidly become much convoluted. The Malpighian body is formed at the inner end of the nephrostomial funnel, and after its formation the funnel separates off, and, according to Hoffmann, ends blindly.

Hoffmann and Selenka‡ lay stress on the fact that the Wolffian tubules develope from behind forwards, appearing first at the extreme

---

* Fürbringer, loc. cit., pp. 54 et seq.
† Hoffmann, loc. cit., pp. 573 and 592.
‡ Selenka, loc. cit., cf. especially Fig. 1.

hinder end of the kidney. They note also, as has been pointed out by Spengel, Fürbringer, and others, that the tubules are at first segmentally arranged, but soon become more numerous than the segments; and that the most anterior tubules, which extend forwards to within two or three segments of the head kidney, differ slightly from the posterior ones, and early undergo degenerative changes.

Sedgwick[*] has disputed some of the above statements, and states very clearly that in the frog the cells from which the Wolffian tubules arise are "at their first appearance independent of the peritoneum, and only secondarily become connected with it." He suggests also that Fürbringer has fallen into error through assuming that the details of formation of the extreme anterior end of the Wolffian body, where the tubules are rudimentary and never attain full development, apply also to the functionally active tubules of the hinder part, or kidney proper.

Our own observations support Sedgwick's statements. We find that in tadpoles of from 10 mm. to 12 mm. length the Wolffian tubules arise as little masses of cells in the mesoblast between the aorta and the archinephric duct, a little distance from the peritoneum, and quite independent of it. These masses of cells are at first segmentally arranged; they are ill-defined groups of spherical or slightly branched cells, which rapidly acquire more definite rod-like shape, then become tubular, and growing outwards meet and open into the archinephric duct, while at their opposite ends Malpighian bodies are formed at a slightly later date. The nephrostomes appear to us to arise as outgrowths from the tubules towards the peritoneum, and the mode of their formation will be described in the next section.

Though we feel fully satisfied as to the origin of the Wolffian tubules in the mesoblast independently of the peritoneal epithelium in tadpoles of *Rana temporaria*, the only species investigated by us, we do not wish to attach much value to the fact from a morphological standpoint. We are rather of opinion that this will prove to be one of the numerous details of development in which the frog does

---

[*] Sedgwick, "On the Early Development of the Anterior Part of the Wolffian Body and Duct in the Chick, together with some remarks on the Excretory System of the Vertebrata," "Quarterly Journal of Microscopical Science," new series, vol. xxi., 1881, pp. 448-450.

not recapitulate its ancestral history correctly, and of which the true history is far more perfectly preserved by the Urodeles. The accounts of the development of the kidneys in Urodeles given by Hoffmann, Fürbringer, and others, are far too precise and consistent to be put aside; and we see no reason whatever for doubting that in Urodeles the Wolffian tubules really arise as peritoneal ingrowths, but it by no means follows that this must apply to the Anura as well. It would be easy to give a long list of developmental details, some of them of considerable importance, in which the frog differs markedly from the newt. It will suffice here to mention the double layered condition of the epiblast in the frog larva, and the consequent modifications in the development of the nervous system and sense organs, the formation of the anus by an independent proctodæal invagination, and the mode of development of the Müllerian duct In these, as in numerous other respects, the frog's development is widely different from that of the newt, while comparison with other Vertebrates shows us further that of the two the Urodele type of development is the simpler and more primitive, the Anuran the more specialised.

2. *The Nephrostomes of the Wolffian Body.*—We propose first to deal with the relations of the nephrostomes in the kidney of the adult frog, and then to consider their condition and mode of development in tadpoles of various ages.

The existence of numerous ciliated funnel-like openings, or nephrostomes, on the ventral surface of the kidney in young frogs, was first discovered by Spengel*, in 1875, by the examination of fresh kidneys by transmitted light.

Independently of Spengel, and almost simultaneously with him, Fritz Meyer† discovered the nephrostomes in the kidneys of adult frogs by staining the peritoneal surface with nitrate of silver. In the kidney of an adult male frog he counted as many as 195 nephrostomes.

In the following year, Spengel published a much fuller account

---

* Spengel, "Wimpertrichter in der Amphibienniere," "Centralblatt f.d. med. Wissenschaft," 1875, No. 2.

† Fritz Meyer, "Beitrag zur Anatomie des Urogenitalsystems der Selachier und Amphibien," "Sitzungsberichte der Naturforschenden Gesellschaft zu Leipzig," 1875, Nos. 2, 3, 4, p. 38.

of the nephrostomes in his well-known and admirable essay on the Urinogenital Organs of Amphibians*. He repeated Meyer's method of treatment with silver nitrate, but obtained the best results by means of chromic acid. He describes the nephrostomes in *Rana, Bufo, Bombinator*, and *Discoglossus*, and gives excellent figures of surface views of the kidneys of *Rana temporaria* and *Discoglossus pictus*, showing the actual arrangement of the openings. The nephrostomes are confined to the part of the kidney covered by peritoneum, i.e., to the ventral surface and inner border; they are arranged along definite lines or tracts, usually following the renal veins, and they are more numerous in the anterior than in the posterior part of the kidney. They are often placed in groups opening into shallow depressions of the kidney surface, and from 200 to 250 can be counted in a single kidney. A single nephrostome may lead into two or three tubules; or a single tubule on reaching the surface may branch and open by two or more nephrostomes.

Spengel notes that although the nephrostomes can be seen with the greatest ease either in surface views or in sections in any plane, yet that he experienced the greatest difficulty in trying to trace them into connection with the kidney tubules. In Cœciliæ and in Urodeles he followed the nephrostomial tubes repeatedly and without difficulty into the necks of the Malpighian bodies; but in none of the genera of Anura examined by him, *Rana, Bufo, Bombinator*, and *Discoglossus*, was he able to trace such a connection. Examination of sections in various planes, injection of the urinary tubules, and teasing of fresh specimens, all alike failed to show any communication. In one single instance, in a transverse section of the kidney of *Bufo cinereus*, did he satisfy himself of the existence of a tubular communication, and that was not between the nephrostomial tubule and the neck of the Malpighian body, but between a ciliated tubule and the fourth section of the urinary tubule, i.e., the collecting tubule. The ciliated tubule was, however, not traced to a nephrostome, and so even this solitary instance is not a proved case.

Spengel, however, considers it probable that the connection in this instance is a real one between a nephrostomial tubule and the fourth

---

* Spengel, "Das Urogenitalsystem der Amphibien," Part i.; "Der Anatomische Bau des Urogenitalsystems," Arbeiten, a.d.; Zool. Zootom., Institut. der Univ. Würzburg, Bd. iii., 1876, pp. 82-89.

part of a urinary tubule, and while recognising the need for further evidence, is inclined, on the strength of this one doubtful instance, to regard such a connection as occurring generally in the Anuran kidney.

The presence of numerous nephrostomes in the kidney of the adult frog, though conclusively proved by Spengel's description, has not yet, in this country, obtained general acceptance; and in the most recent work on the anatomy of the frog, the English translation of Ecker's well-known book, the editor states that after careful examination of sections, teased specimens, and fresh and injected kidneys, he was unable to find any trace of the nephrostomes.*

Under these circumstances, we have thought it well to examine carefully the structure of the kidney in adult and in young frogs, as well as in tadpoles.

In the kidney of the adult frog we find the nephrostomes present in large numbers, exactly as Spengel describes them. Fig. 7 represents a transverse section through the kidney of an adult male frog, taken not very far from the anterior end. The dorsal half of the kidney is formed mainly by the large collecting tubes K T. In the ventral half lie the narrower and much convoluted "third sections" of the urinary tubules, K R, and also the Malpighian bodies, K G. On the ventral surface, which is covered with peritoneum, three nephrostomes are shown, each commencing with a mouth, E A, opening into the body cavity, and surrounded by a slightly raised lip, which leads into the short nephrostomial tubule, E. The walls of the tubule are composed of a single layer of small cubical epithelial cells with pigmented walls, each cell bearing a long flagellum, which lies in the tubule with its free end directed inwards. Round the mouth of the tubule the flagella form a ring projecting into the cœlom.

We have found no difficulty in seeing these nephrostomes in any good series of sections of frog's kidney, and entirely fail to understand why they have been overlooked by other observers. They are not present in every section, but an examination of half a dozen consecutive sections is certain to show one or more. Fig. 7 is not drawn from a single section, but from four consecutive ones, as a

* "The Anatomy of the Frog," by Dr. Alexander Ecker; translated with numerous annotations and additions by Dr. J. Haslam. Oxford, 1889, p. 336.

single section rarely shows the whole length of a nephrostomial tubule.

By snipping off with scissors small portions from the ventral surface of the kidney of a freshly pithed frog, and examining in normal saline solution, we have seen without difficulty the nephrostomes in a living condition, and the flagella in active motion. It is noteworthy that in such preparations the current is an outward one through the nephrostome into the body cavity; and also that the flagella along the whole length of the nephrostomial tubule have their free ends directed outwards towards the body cavity, and not, as in all our sections of hardened kidneys, inwards.

A further point of great interest in regard to the nephrostomial tubules of the adult frog, concerns the mode of termination of their inner ends.

Nussbaum, who had previously supported Spengel's view that the nephrostomial tubules in adult Anura open into the fourth section of the urinary tubules, corrected himself in 1880,[*] and stated as the result of further investigations that the nephrostomial tubules open in young larval Anura into the neck of the Malpighian bodies, as they do throughout life in the Urodeles; but that in adult Anura, at any rate in the genera *Rana*, *Bufo*, and *Bombinator*, the inner end of the nephrostomial tubule, though it may lie in contact with the neck of a Malpighian body, actually ends by opening into the veins of the ventral portion of the kidney.

In a subsequent paper, Nussbaum[†] gives the results of a further investigation by himself and by his pupil Wichmann, which confirms in all respects his previous account. He also gives a series of excellent figures showing the openings of the nephrostomial tubules into the renal veins in *Rana fusca*, *Rana esculenta*, *Bufo calamita*, and *Alytes obstetricans;* and showing also the nephrostomial tubule opening into the neck of the Malpighian body in young larvæ of *Rana fusca*, but separating from the Malpighian body and opening into the renal vein in an older larva of the same species.

---

[*] Nussbaum, "Ueber die Endigung der Wimpertrichter in der Niere der Anuren," "Zoologischer Anzeiger," iii., 1880, pp. 514-517.

[†] Nussbaum, "Ueber den Bau und Thätigkeit der Drüsen," Part v.; "Zur Kenntniss der Nierenorgane," "Archiv für Mikroskopische Anatomie," xxvii., 1886, pp. 466-469, and plate 23.

Nussbaum's method of proceeding is as follows: He first kills the frog with chloroform, then injects into the body cavity carmine rubbed up finely in normal salt solution; the wound is closed at once, which is conveniently effected by twisting a ligature round a couple of needles passed through the body wall close to the lips of the wound, and the frog is at once placed in Müller's fluid. After about three hours the body cavity is cut open so as to allow the Müller's fluid free access to the kidneys, and the frog is then left in the fluid for twelve to twenty-four hours longer, when the kidneys are removed and examined either by teasing or by sections. The flagella lining the nephrostomial tubules continue their movements for some time after death of the frog, and in this way the carmine is drawn through the nephrostomes. In sections or teased preparations, the carmine particles are found in the nephrostomial tubules, and also in the blood vessels around them, but in no part of the urinary tubules.

We have repeated Nussbaum's experiments with complete success, and have also been able by examination of sections of kidneys prepared in the ordinary way to demonstrate conclusively the correctness of his description.

In Fig. 7, the middle one of the three nephrostomes figured is shown ending with an open mouth into the renal veins; and in Fig. 4, one of the nephrostomial tubules of an adult frog is drawn on a larger scale. The tubule commences with an open, somewhat funnel-shaped mouth, E A; the tubule itself runs inwards, not vertically to the surface of the kidney, but almost parallel to it, a condition we have found to be almost invariable (*cf.* Fig. 6). The tubule is somewhat conical in shape; its inner end is narrow, but opens by a distinct aperture into the vein surrounding the tubule. The wall of the tubule consists of a single layer of epithelial cells, columnar round the peritoneal mouth or nephrostome, cubical or slightly flattened along the rest of its length. Round the peritoneal opening, the columnar or cubical cells of the tubule extend outwards a short distance, forming a lip to the nephrostome, and then pass suddenly into the squamous cells of the peritoneum. Each of the epithelial cells of the tubule bears a single long flagellum; round the lip of the nephrostome these flagella project into the body cavity, but their direction is inwards towards the nephrostome. In the tubule itself the flagella, which are longer than the diameter of the tubule, are directed inwards, and through the terminal aperture of the tubule they project as a tuft some distance into the vein. In all our sections of hardened kidneys the flagella are directed inwards; but in separated portions of fresh kidneys, as noticed above, we have found them pointing outwards.

About the existence of these openings into the veins we have no doubt whatever; we have seen them with perfect distinctness in a large number of specimens prepared in different ways, and from different frogs. In specimens injected with carmine, after Nussbaum's method, the particles of carmine may be seen in the nephrostomial tubules, and passing through their open mouths into the veins, while no carmine is found in any other portion of the urinary tubules; but sections prepared in the ordinary way show the communication quite as clearly. When the blood corpuscles are numerous, the tuft of flagella may be seen projecting from the mouth of the tubule, between and among the corpuscles.

In his first paper, Nussbaum speaks of the veins into which the nephrostomial tubules open as the portal veins; but in his second paper he calls them the branches of the posterior vena cava, *i.e.*, the renal veins. We have satisfied ourselves that the latter determination is the correct one. The veins into which the tubules open lie along the ventral surface of the kidney, and can readily be traced into the vena cava.

In Fig. 3 we have drawn one of the nephrostomial tubules from the kidney of a 40 mm. tadpole (*cf.* Figs. 1 and 9). The relations are seen to be precisely the same as in the adult frog, except that the tubule is rather shorter, and the lip round its peritoneal opening less prominent. The tubule has no relation, except one of apposition, with the urinary tubules, and opens by a conspicuous aperture, E B, through which the tuft of flagella projects, into the renal vein, V R.

Fig. 2 shows identical relations as seen in the kidney of a 21 mm. tadpole (*cf.* Fig. 6).

Earlier than 21 mm., our observations are less complete. In 17 mm. tadpoles the nephrostomial tubules are present, and open into the body cavity at their outer ends. The inner ends of the tubules appear to us to end blindly, but it is not easy to determine the point satisfactorily.

Concerning the mode of formation of the nephrostomial tubules, our observations are also incomplete. The tubules appear to us to arise as solid rods of cells in the mesoblast, continuous with the strings of cells which give rise to the Wolffian tubules. These soon acquire connection with the peritoneum, the peritoneal epithelium however taking no part in their formation, and the whole of the epithelium of the tubule being derived from the solid rod of mesoblast cells. The lumen of the tubule is formed as a split between the inner ends of the cells of the rod, and not by invagination of the peritoneum.

We have not ourselves seen at any stage a tubular communication between the nephrostome and the neck of the Malpighian body, or indeed with any other part of a urinary tubule. Nussbaum's description and figures of the early communication with the neck of the Malpighian body are, however, very precise, and they have been confirmed by Hoffmann.* Bearing in mind also that Spengel and Nussbaum's descriptions show conclusively that in Urodeles this connection between the nephrostomial tubule and the neck of the Malpighian body is not only present in the larva, but persists throughout life, we are inclined to believe that it really occurs in young Anuran tadpoles, though it can only be present for a very short period, and may perhaps not be acquired in all cases. If this view is correct, we have in the relations of the nephrostomial tubules another instance of a curiously specialised condition in Anura, derived from a more primitive condition which is still retained by the Urodeles.

III. The Development of the Fat-bodies.

Von Wittich† first described the fat-bodies of Amphibians as formed by fatty degeneration of the anterior ends of the genital ridges.

Giles‡ has recently questioned the accuracy of this determination,

---

* Hoffmann, loc. cit., pp. 593, 594. Hoffmann maintains that in the adult Anura the nephrostomial tubules end blindly.

† Von Wittich, "Beiträge zur Morphologischen und Histologischen Entwickelung der Harn, und Geschlechtswerkzeuge der nackten Amphibien;" "Zeitschrift f. wiss. Zoologie," Bd. iv., 1853, pp. 148, 149.

‡ Giles, "Development of the Fat-bodies in Rana temporaria;" "Quarterly Journal of Microscopical Science," vol. xxix., 1888, pp. 133-142, reprinted in the present volume.

holding that the fat-bodies are really formed by fatty degeneration of the head kidneys, and that their connection with the genital organs is a secondarily acquired one.

We have investigated the point, and find that von Wittich's account is correct.

Fig. 1 represents, as already noticed, a 40 mm. tadpole dissected from the ventral surface. The condition of the urinary organs has been sufficiently described above. The genital organs, G R, are present as a pair of narrow ridges lying along the sides of the aorta and the inner borders of the Wolffian bodies. The anterior part of each ridge, G F, is separated by a slight constriction from the posterior part, and is notched at its free edge into a few small, irregular lobes. This anterior part of the genital ridge is the commencing fat-body; it is, as von Wittich correctly described, almost invariably larger on the left side than the right.

Figs. 8 and 9 represent transverse sections passing respectively through the fat-bodies and through the genital organs of a tadpole of the same age as Fig. 1; the levels at which the sections are taken being indicated by the lines aa, bb, in Fig. 1.

The fat-bodies (Fig. 8, G F), which along the greater part of their length lie alongside the archinephric ducts, but in front of the Wolffian bodies, are seen to have a very similar appearance, and precisely the same relations, as the genital organs in Fig. 9, G R; while Fig. 1 shows how far they are really removed from the head kidneys.

At this period the fat-body consists of a solid mass of closely packed cells, in which the deposition of fat is only just commencing.

The condition in a tailed frog is shown in Fig. 5, in which the head kidneys, N A, have almost completely disappeared; while the fat-bodies, G F, have increased greatly in size, and are produced at their free ventral borders into finger-like lobes. The fat-body of the left side is still markedly larger than that of the right side.

Giles' error appears to have arisen from his removing the kidneys and genital organs before cutting them into sections, and so overlooking the real position and relations of the head kidneys in the tailed frog, and mistaking for them the degenerating tubules of the anterior end of the Wolffian body, in which true fatty degeneration occurs as he correctly describes; while Figs. 8 and 9 show how easy it would be for sagittal sections, such as he figures, to give false impressions concerning the relations of the fat-bodies to the kidneys.

IV. Summary of Results.

Our own observations are confined to tadpoles and adults of *Rana temporaria*.

1. The head kidney and its duct are mesoblastic structures: the epiblast takes no share in their formation; but the cloacal openings of the ducts are formed partly by outgrowth from the gut.
2. The head kidney is a large and complex organ during early larval life, but begins to degenerate in tadpoles of about 20 mm. length.
3. The first stage in the degeneration of the head kidney consists in great and irregular dilatation of the tubules, accompanied by destruction of their epithelial lining, and apparently due to blocking of the archinephric duct.
4. The head kidney ultimately completely disappears; all three nephrostomes close up, and atrophy.
5. The peritoneal opening of the oviduct is a new formation, and not a persistent nephrostome; it is produced in a somewhat complicated fashion.
6. The tubules of the Wolffian body begin to form shortly before the head kidney degenerates; they develope from behind forwards, and are at first segmentally arranged.
7. The Wolffian tubules are formed from the mesoblast between the aorta and the archinephric ducts, and arise independently of the peritoneum.
8. Nephrostomes are formed at an early period, and open into the body cavity. There is reason for thinking that they at first communicate with the necks of the Malpighian bodies; but they very early lose all connection with the urinary tubules, and in the later larval stages, as well as in the adult, they open directly into the renal veins.
9. The fat-bodies are formed from the anterior ends of the genital ridges.

DESCRIPTION OF PLATE X.

Illustrating Professor Marshall and Mr. Bles' Paper "On the Development of the Kidneys and Fat-bodies in the Frog."

Alphabetical List of Reference Letters for all the Figures.

A	Dorsal aorta	M	Spinal cord
A P	Pulmonary artery	M G	Spinal ganglion
B G	Portion of body cavity surrounding the glomerulus	M N	Spinal nerve
		M N 2	Brachial nerve
C	Red blood-corpuscle	M O	Dorsal root of spinal nerve
D	Mouth	M P	Ventral root of spinal nerve
D E	Œsophagus	M S	Sympathetic nerve cord
D I	Intestine	N	Notochord
D L	Lower lip	N A	Head kidney
D O	Cloacal aperture	N B	Tubule of head kidney
E	Nephrostomial tubule	N C	Enlarged degenerating tubule of head kidney
E A	Peritoneal opening of nephrostomial tubule	N D	Archinephric duct
E B	Opening of nephrostomial tubule into blood-vessel	N G	Glomerulus of head kidney
		N T	Nephrostome of head kidney
E F	Flagellum	O	Opercular cavity
G F	Fat-body	O G	Internal gill
G R	Genital ridge	P	Peritoneum
H	Ventricle of heart	P M	Mesentery
K	Wolffian body	P T	Mesorchium
K B	Bowman's capsule	R	Recessus labyrinthi
K G	Glomerulus	S	Septum formed by fusion of lung with head kidney, partially boxing in the glomerulus
K M	Malpighian body		
K N	Neck of Malpighian body		
K R	Tubule of Wolffian body, convoluted portion		
		T	Myotome
K T	Collecting tubule	V C	Posterior vena cava
K U	Ureter	V P	Pulmonary vein
K V	Opening of ureter into cloaca	V R	Renal vein
L	Lung	V S	Renal portal vein
L F	Fore limb	V N	Posterior cardinal sinus
L H	Hind limb	Y	Lymph space

Fig. 1. A 40 mm. tadpole dissected from the ventral surface to show the condition of the head kidneys, Wolffian bodies, fat-bodies, and genital ridges. The dotted lines $a\,a$ and $b\,b$ show the lines along which the sections in Figs. 8 and 9 are taken respectively : ($\times$ 5).

Fig. 2. Part of a transverse section through the Wolffian body of a 21 mm. tadpole, showing the whole length of one of the nephrostomial tubules, with its openings into the cœlom and into the renal vein : ($\times$ 120).

Fig. 3. Part of a transverse section through the Wolffian body of a 40 mm. tadpole (*cf.* Figs. 1 and 9). The figure shows the whole length of one of the nephrostomial tubules, with its openings into the cœlom and into the renal vein : ($\times$ 120).

Fig. 4. Part of a transverse section through the kidney of a frog at the commencement of the second year (*cf.* Fig. 7). The figure shows the whole length of one of the nephrostomial tubules, with its openings into the cœlom and into the renal vein : ($\times$ 120).

Fig. 5. A tailed frog during the metamorphosis ; dissected from the ventral surface to show the condition of the head kidneys, Wolffian bodies, fat-bodies, and genital ridges : ($\times$ 5).

Fig. 6. The dorsal half of a transverse section through a 23 mm. tadpole at the level of the fore limbs and the second spinal nerve. The figure shows the head kidney with the tubules enlarged and degenerating ; also the boxing in of the glomerulus on each side by fusion of the lung with the head kidney : ($\times$ 22).

Fig. 7. A transverse section through the kidney of an adult male frog. The figure has been constructed from four consecutive sections. The dorsal surface is turned upwards in the figure, and the ventral surface downwards ; the inner border of the kidney is to the left, the outer border to the right side of the figure. Three nephrostomial tubules are shown ; in all three cases the peritoneal openings of these tubules are drawn, and in the middle one of the three the opening into the renal vein is shown as well : ($\times$ 45).

Fig. 8. The dorsal portion of a transverse section through a 40 mm. male tadpole, passing through the fat-bodies and the archinephric ducts. The section is taken at the level indicated by the line aa in Fig. 1 : (× 28).

Fig. 9. The dorsal portion of a transverse section through the same tadpole as Fig. 8, passing through the genital ridges and the Wolffian bodies; on each side a nephrostomial tubule is shown. The section is taken along the line bb in Fig. 1 : (× 28).

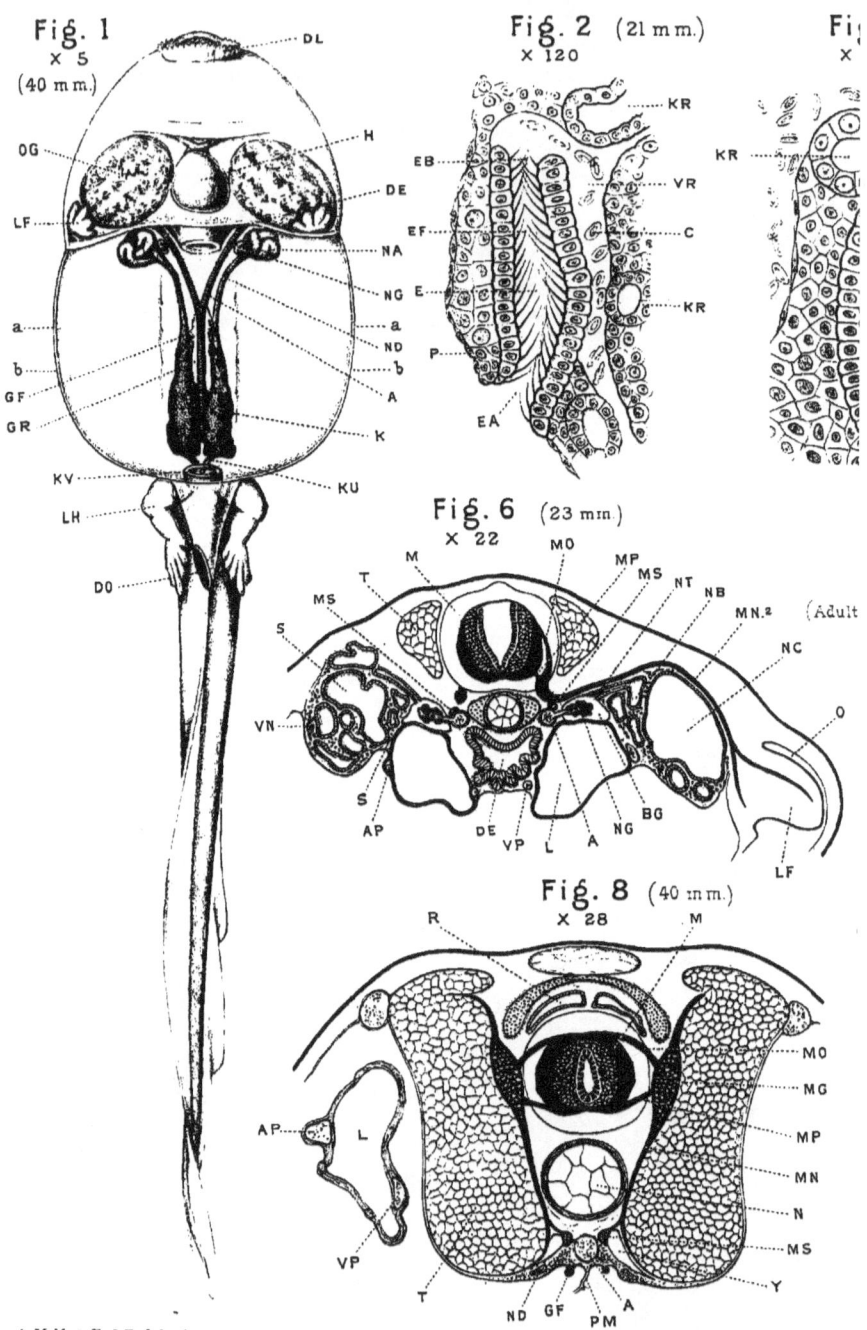

Plate X.

(40 mm.)

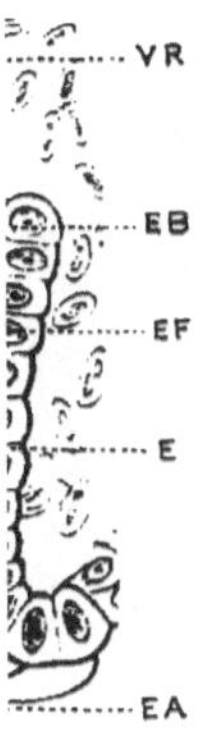

Fig. 4
x 120
(Adult frog)

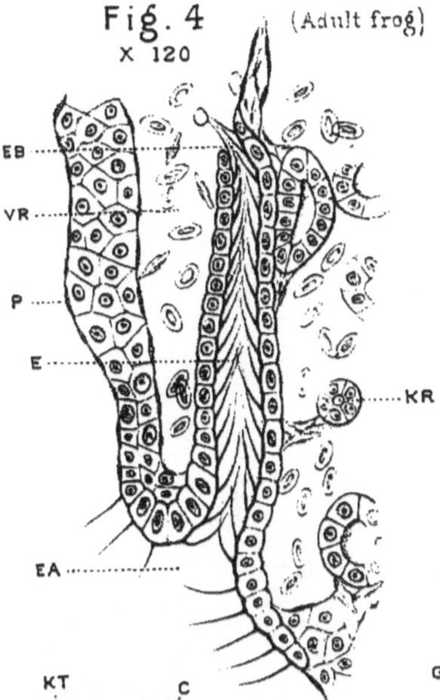

7

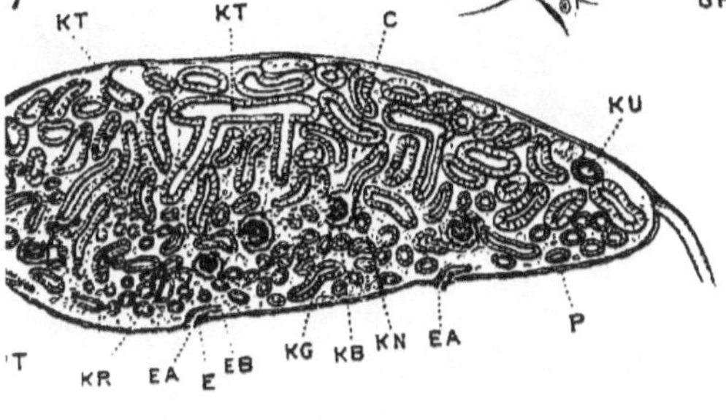

. 9

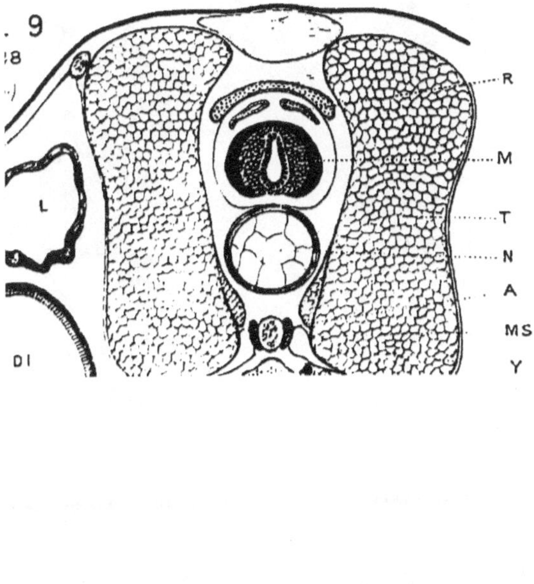

Fig. 5
x 5
(Tailed frog)

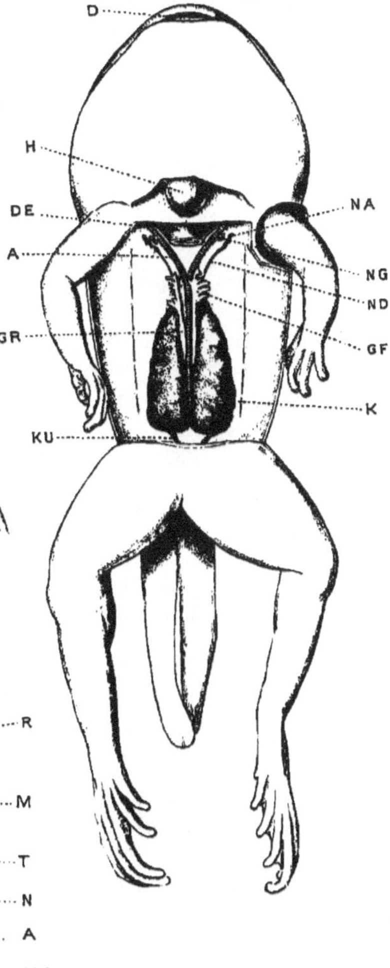

THE DEVELOPMENT OF THE EAR AND ACCESSORY ORGANS IN THE COMMON FROG.

By Francis Villy, *Owens College, Manchester.*

[With Plates XI. and XII.]

The development of the auditory organ of Vertebrates is an embryological subject that has hitherto received very little attention. The ear has been studied more completely in the higher than in the lower types, and mammals have received an amount of notice in this respect altogether out of proportion to the value of the evidence that their highly modified development may reasonably be expected to yield.

Under these circumstances I have, at Professor Marshall's suggestion, undertaken an investigation of the development of the ear of the common frog *(R. temporaria)*, including the associated organs, from the earliest stages to the permanent form.

The frog's ear has been chosen as a subject for the reason that it probably represents an important stage, more complicated than the simple internal ear of fishes, and lower than the condition found in higher Vertebrates. At the same time the frog may be taken as a type of the lowest class possessing a Eustachian tube and an auditory ossicle, and as such may reasonably be expected to throw some light on the history of these organs during its metamorphosis from tadpole to frog, from typically aquatic to typically terrestrial structure. Besides these advantages possessed by the frog as an animal suitable for study, embryos and tadpoles may readily be obtained at any desired stage, from the egg to the adult.

The investigations here recorded have been conducted almost entirely by means of transverse sections, longitudinal sections being examined when points of special difficulty were encountered.

Before passing on to the immediate subject of this paper I wish to take this opportunity of thanking Professor Marshall for the help that he has so readily given me, without which my work would have been lengthened, and rendered much more laborious than it has proved to be.

Early Stages.—The earliest stage at which the auditory involution can be readily recognised occurs soon after the closure of the neural groove in the fore part of the embryo. It will be seen from Fig. 2 that the invagination, though shallow, is perfectly distinct, the epithelium lining it being somewhat sharply marked off from the surrounding epiblast by the columnar character of its constituent cells. The invagination only concerns the deeper or nervous layer of epiblast, and is placed in the longitudinal groove separating the lower part of the embryo, distended with yolk, from the neural surface, and a little above the level of the notochord. The edge of the pit is roughly circular, and as yet the involution is regular, being deepest in the centre, and shallowing in all directions towards the edge.

Although I have taken this stage as the earliest in the development of the ear, the auditory epithelium may be recognised with more or less certainty before the neural groove closes in. In these very early stages the deeper layer of epiblast is somewhat thickened in the region where the ear is formed later, and the auditory nerve may already be traced into continuity with it.

As development proceeds the involution is more rapid at the dorsal than at the ventral part, so that in section it comes to have the appearance shown in Fig. 2. After this has taken place the ventral part in turn grows rapidly, pushing its way inwards (see Fig. 3), and as this process is going on the lips of the pit grow towards each other and coalesce, thus closing the aperture completely. Soon afterwards the newly formed outer wall of the vesicle separates from the deeper layer of epiblast, which remains near the surface. I am not certain that the mouth of the auditory sac is closed precisely in this manner, for the nervous layer of epiblast is so ill defined at this time and place that I have thought it more probable that the outer wall of the vesicle is formed simply by the cells at the edges of the pit proliferating, and closing the mouth by forming a plate across it.

Whatever the real mode may be, the vesicle is pyriform when first closed, the dorsal part being the narrower. This narrow part ultimately becomes the recessus labyrinthi. The outer wall of the vesicle, being the latest formed, consists at first of cells indistinct in outline, but soon they assume, more or less completely, the columnar character of the opposite wall, though the two parts never agree exactly, as the older contains more pigment than the younger. It is from the old pigmented parts that the sensory tracts of the adult ear are formed.

Very soon after the auditory sac is completely closed, irregular mesoblast cells make their way into the space between it and the external epiblast, and by this means the organ is removed farther and farther from the surface.

The whole of the changes mentioned above take place before the tadpole character is assumed, and the invaginated vesicle, such as is here described, may be found in just hatched larvæ of about 4 mm. in length. No great advance in complication is to be noted until the sacculus is marked out and the semicircular canals begin to form. This does not take place until the larva is a well-developed tadpole of 11 mm. or 12 mm. in length.

I propose now to give a short description of the organ as present in tadpoles of this latter size, and then to describe the changes that take place in the various parts during the older stages, and the accessory structures that arise in connection with the ear at a later period.

The Invaginated Vesicle.—In tadpoles of this size, *i.e.*, up to about 11 mm., the auditory sac is spherical in general shape, its wall consisting of a single layer of cells. At one point in the centre of the dorsal surface the regular contour is broken by a fold of the wall projecting slightly upwards, so that, as already mentioned, the vesicle is pyriform in outline (see Fig. 4). This fold of the upper surface is at first only slightly marked off from the rest of the vesicle, but gradually it becomes narrower, and at the same time is pushed towards the inner side by irregular growth of the wall, so that it comes to lie in the position shown in Fig. 5. This fold is the first rudiment of the recessus labyrinthi. As mentioned above, it can be recognised from the moment the wall of the sac is completed, and it results from the mode of involution, but it is not a stalk connecting the vesicle with the exterior at any stage. The typical Vertebrate ear retains its opening to the exterior for some time by means of such a recessus labyrinthi, and the duct may even be present in the adult, putting the endolymph in communication with the surrounding medium (Elasmobranchs); but when the ear-fold concerns the inner epiblast alone no such communication takes place at all. In the case of the frog we should expect the recessus to be the last part of the vesicle to retain its connection with the external skin, but this is not so. This irregularity may perhaps be explained by the double-layered condition of the epiblast, which condition is most probably a secondarily acquired one.

The epithelium forming the vesicle is more or less columnar at all points. The older part, *i.e.*, that forming the wall next the brain, the floor and the recessus, consists of cells differing from the remainder in being more elongated. This is most marked at the lower inner angle, where the cells are deeply pigmented. This tract of elongated cells extends upwards and outwards at the anterior end, especially in the later stages of this period.

The auditory nerve runs from the inner wall, with the greater part of which it is connected, and reaches the brain just behind the recessus. The part applied to the vesicle is expanded into a ganglion of some size, whilst the nerve connecting it with the brain consists at this stage of a fairly thick strand of cells. The whole is pigmented. During this period, or at least nearly the whole of it, until the tadpole is about 11 mm. in length, the auditory nerve is absolutely continuous with the facial nerve. After this stage the ear becomes too complicated to describe clearly as a whole; I propose, therefore, to take the various parts one by one and follow them up to the adult condition in this order :—

    1. The Differentiation of the parts of the Internal Ear.
        a. The Semicircular Canals.
        b. The Sacculus.
        c. The Cochlea.
        d. The Epithelium of the Auditory Vestibule.
        e. The Recessus Labyrinthi.
        f. The Perilymphatic Spaces.

2. The Hyomandibular Cleft, Eustachian Tube, and Tympanic Cavity.

3. The Annular Cartilage, the Stapes, and the Columella.

a. The Semicircular Canals.—These canals are essentially parts of the auditory vesicle which have become separated from the main part by septa which are incomplete, inasmuch as each canal opens into the vesicle at each end of the septum. Each septum originally consists of two entirely distinct folds which grow together from opposite walls of the vesicle, and by their coalescence cut off the canals except at their ends. The anterior vertical and the horizontal canals develope simultaneously in the frog, whilst the posterior vertical is distinctly later in formation.

The first indication of the canals is afforded at about 11 mm., by folds of the walls of the vesicles projecting internally. These folds are at first purely epiblastic; but very soon irregular mesoblastic cells migrate into the space included in each double fold. The ridges so formed are placed one at the anterior end, in the upper and inner part of the vesicle, and one opposite this and projecting towards it from the upper and outer part, and running back nearly to the hinder part of the ear. In its hinder half this latter projects ventrally instead of horizontally; in fact, it is almost completely divided into two parts, the anterior corresponding to the first ridge mentioned, and with it going to form the septum of the anterior vertical canal, whilst the posterior portion lies opposite to a third fold projecting upwards from the ventral part of the vesicle, these two constituting the septum of the horizontal canal.

The two folds of each pair grow towards each other till they meet and coalesce for a considerable distance. At the stage now reached the anterior vertical and horizontal canals are already established. Figs. 7 and 8 show two stages in the formation of the canals as here described. In Fig. 7 the septa have not yet been completed, whilst in Fig. 8 the constituent folds have met and blended. It should be noted, however, that these two sections do not exactly correspond, for they cut the head at different levels. The septa formed in this way rapidly thicken and elongate, thus lengthening the canals, this lengthening keeping pace with the general growth of the vesicle during this period.

Whilst these changes have been going on the posterior vertical canal has been forming. The first signs of its presence are apparent some time after the ridges forming the septa of the other canals have appeared, but before the coalescence of these ridges. It is formed exactly as the other canals are; the outgrowths developing, one on the septum between the sacculus and utriculus hereafter to be described, and situated on its outer edge directed upwards and inwards; the other is on the inner wall of the vesicle, more than half way up it and opposite to the first. This is shown in Fig. 8. These folds coalesce after the time of cutting off of the two other canals. In the adult the posterior vertical canal curves over the horizontal canal and opens into the utriculus below it. This is brought about by the formation of the hinder ampulla, which by its growth causes an apparent inversion of these parts. The part of the utriculus originally above the horizontal canal developes into the ampulla, and in so doing bulges out and encircles the canal.

The two vertical canals are at first entirely distinct, but as they lengthen they approach each other and meet. In this way a common length is formed, short at first, but gradually elongating, and through it both canals open into the utriculus.

With this stage the adult relations of the canals are attained, further change being limited to their change in size, and the growth of mesoblast between them and the rest of the auditory vesicle, whereby they are removed farther and farther outwards and attain their characteristic curves. The mesoblast concerned consists at first of irregular scattered cells, but as the cartilaginous auditory capsule developes the cartilage extends between the canals, and in tadpoles of about 20 mm. it entirely surrounds them.

With regard to the relative times of formation of the canals, the anterior vertical and horizontal are formed simultaneously, but the posterior vertical is only just commenced when the other two are nearly cut off, and there are stages in which the posterior vertical canal is not fully separated by the blending of the folds constituting its septum, whilst the other two have been formed some time. It is noteworthy that of the sensory epithelia of the ampullæ, that of the posterior is the first to be specialised, though the corresponding canal is the last to develope. This will be more fully described and an explanation offered in dealing with the differentiation of the sensory epithelium of the ear.

It is usually stated that the two vertical canals in the higher Vertebrates are the first to develope, and that they correspond to the two canals of the Cyclostomata. From this it is argued that the Cyclostome ear is more primitive than others, a stage representing their condition being passed through in the development of higher forms. It will be seen from the above that this doctrine is very much shaken by the results arrived at in this paper.

The ampullæ, concerning the formation of which nothing has as yet been said, are developed by the constricting off of parts of the vesicle at the ends of the canals. The sensory epithelium of each ampulla is present some time before the ampulla itself, and this sensory epithelium is situated at the appropriate end of its canal. As the ampullæ are formed, they therefore include their special epithelial tracts from the very first.

b. The Sacculus.—In tadpoles of about 11 mm. in length, just before any signs of the semicircular canals appear, the internal ear is complicated by the formation of a septum running obliquely across its hinder part from the upper and outer to the inner and lower angle. This septum seems to grow forwards from the posterior wall of the vesicle, first appearing as a ridge and gradually broadening till it divides the posterior part of the vesicle into a larger upper and inner and a smaller lower and outer part (see Fig. 8). These two divisions of the vesicle communicate with the rest of the cavity in front, and therefore indirectly with each other. This septum is the first sign of a division into utriculus and sacculus. The upper portion of the vesicle becomes the hinder part of the utriculus in the adult, that is chiefly the posterior ampulla. The septum under consideration is at first confined to the posterior part of the vesicle, and remains so for some time; but later on it extends all round the vesicle, and growing inwards completes the division into utriculus and sacculus. This division is not quite complete, for the septum remains perforated at its centre, thus leaving the utriculus and sacculus in communication. The septum consists of a double epiblastic fold for the most part. It attains the proportions of the adult in tadpoles of about 20 mm.

Although the sacculus is formed in part merely by its separation from the rest of the vesicle by an incomplete septum, still it manifests active growth, and the great downwardly projecting pouch of the sacculus is mainly formed as an outgrowth of the vesicle. This is plainly seen just after the completion of the semicircular canals. At this time a slight bulging of the floor of the vesicle may be noticed near the anterior end of the horizontal canal. This forms the pouch of the sacculus spoken of above. It enlarges rapidly, and soon loses its distinct character, as its opening increases more quickly than the remainder, so that it never becomes deeply constricted off.

c. The Cochlea.—Under this heading I include the three pouches of the sacculus described by Hasse* and Retzius† as the lagena, the pars neglecta, and the pars basilaris. There is very little of importance to note concerning them except the relative times at which they appear. The lagena is the hindermost and largest of the three. It first appears as a small pouch at the lower and inner angle of the vesicle, opening into it at about the level of the anterior aperture of the posterior vertical canal (see Fig. 8). It early acquires considerable size, and extends back some distance. As the perilymphatic spaces form, as described below, the inner wall of this pouch becomes thin and abuts against one of these canals, the part of the pouch concerned presenting a flattened appearance. The remaining parts of the cochlea develope simply as bulgings of the walls of the sacculus. The first of these to form is the anterior and upper one, which is situated close beneath the mouth of the recessus. This is called the pars neglecta. The pars basilaris is therefore the last part to appear.

d. The Epithelium of the Auditory Vestibule.—In this section I propose to trace the formation of the patches of sensory epithelium in the various parts of the ear.

When the lining of the invaginated auditory vesicle begins to differentiate, cells of two kinds may be distinguished,—columnar and pavement. The columnar cells are disposed in a tract running the length of the floor of the vesicle, and especially well marked at the inner angle. This tract extends up the inner wall for a certain distance, and in the fore part it runs up the outer wall. The remainder of the vesicle is formed of a flattened epithelium. The columnar portion of the epithelium, however, divides and gives rise

---

* "Zeitschr. f. wiss. Zoologie," xviii., " Das Gehörorgan der Frösche."
† "D. Gehörorgan d. Wirbelthiere."

to all the sensory patches present in the adult ear, although at the time under consideration it is continuous.

The sensory patches of the ear are eight in number. Three of these are situated in the ampullæ, three in the cochlea, and of the remaining two one is on the floor of the anterior part of the utriculus, and one on the inner and lower part of the sacculus.

The first division of the columnar epithelium takes place at the hinder end of the vesicle in tadpoles of about 11 mm. It occurs on the formation of the oblique septum, which I have described as marking the first trace of division of the vesicle into utriculus and sacculus. The small sensory patch thus separated lies in the hinder end of the utriculus, that is, where the posterior ampulla is formed later, so that the ampulla as it developes includes this epithelial patch.

The fact that this epithelial tract is the first to develope might be taken to show that though the posterior vertical canal actually forms later than the others, nevertheless it is indicated earlier than these are. This early specialisation may, however, possibly be explained as a convenience due to the out-of-the-way position occupied by the organ concerned and to the early formation of the septum between the utriculus and sacculus, which in its growth plays an important part in separating this patch from the rest. Besides, I do not think that the time of specialisation of the sensory tracts of the ampullæ is a safe guide to the relative times of evolution of the canals, as there is no rule of development common to all three. If the ampullæ were exactly homologous structures, and their development had not been interfered with, they would form in the same way; but as they do not do so, it is safer to assume that the original course of development has been modified than that the ampullæ are not homologous. This would seem to show that the order of the evolution of these epithelial patches has not been preserved.

It is interesting to note that just as the anterior vertical and horizontal canals form close together, and are even united from the first, so the epithelium of their ampullæ developes from a common rudiment, which becomes separated from the main mass soon after the posterior tract already mentioned. This rudiment, common to the two ampullæ, is situated on the outer wall at the anterior end of the vesicle, near the point where the ampullæ of the two canals, with which it is destined to be associated, will be placed. A short time after the canals have been established this patch divides into two, of which one goes to form the sensory epithelium of each ampulla.

The sensory epithelial areas of the cochlea and sacculus remain continuous with each other for some time. When the sacculus and lagena form, it follows from their position that both are lined by columnar epithelium, the former, however, only in part. The rapid growth of the sacculus necessitates the drawing out of the sensory lining of the floor of the vesicle, so that one end is now situated in the anterior part of the vesicle, and from this point it extends backwards to the end of the single cochlea pouch present, *i.e.*, the lagena. The first part of the thickened epithelium to be separated from the rest lies on the under surface of the lip of the recessus, and when the pars neglecta of the cochlea developes it necessarily includes it from its formation at the same point. The next sensory tract that differentiates is separated in the position where the remaining pouch of the cochlea (pars basilaris) will ultimately be, and forms the greater part of the wall pushed out in the development of this organ. The lining of the first formed division of the cochlea, *i.e.*, the lagena, is the last to become distinct from the saccular epithelium. This fact may fairly be taken to support my view that the epithelial patches do not develope in the order in which they were evolved.

I believe that it is of some importance to recognise that the outlying sensory tracts are individualised before the parts to which they are destined to belong. The ampullæ and cochlear pouches are provided for in this way, all except the lagena; but this apparent exception does not really contradict the rule enunciated above, as the pouch under consideration is from the first placed in such a position that unless some secondary change were to take place it must be lined by sensory cells as soon as it is formed. The lagena appears early, when the canals have only just been established and when the epithelium of the ampullæ is being separated, and it arises in the position where the columnar epithelium is best marked. Thus it resembles the other parts to which the rule applies in possessing sensory cells from the first, whilst it differs from them in not having a sense-organ distinct from that destined for other parts. The ampullæ and the remaining two cochlear pouches may be recognised by their thickened epithelium before they can be distinguished by their shape.

The last division which takes place in the sensory tract separates it into two parts, one of which lies in the utriculus and one in the sacculus. It does not take place for some time after the others, and even in the adult these two parts are not far removed from each other.

e. The Recessus Labyrinthi.—The existence of this peculiar organ, the anatomical relations of which have been described by Hasse,* has been overlooked in the Anura. The animals that I have examined *(Rana temporaria, Bufo vulgaris,* and *Dactylethra larva)* all possess very similar structures. Taking the adult frog as a type, the relations of this part of the auditory sac may be briefly noted here. A narrow duct leads from the sacculus near its opening into the utriculus, and, running up the inner surface of the vestibule, perforates the skull wall. Within the skull it is dilated to form a large, thin-walled, vascular sac, which extends some distance both in front of and behind the duct. The sacs of the two sides are connected by bands of spongy tissue above the cerebellum, and below the brain just behind the pituitary body.

The walls are vascular, and Hasse has assigned to this part of the ear the function of supplying the endolymph. It cannot be sensory, for at no period are its walls supplied with nerves in sufficient quantity, and its histological structure will not admit of such an interpretation. It would seem to have some function of importance in the adult, as it steadily increases in size during the growth of the tadpole, and it is after the tadpole stage is passed that this increase in size becomes most rapid and the blood-supply most copious.

The relations of this organ at different stages are shown in Figs. 4, 5, 7, 9, 17, and 18. Its mode of origin and early stages have already been described, and until the semicircular canals are formed little is to be noted except a general growth in size, accompanied by a movement towards the brain, so that it comes to lie in close contact with this organ. As the distal part comes close to the brain, it begins to expand and its duct narrows; at the same time the upper lip of the duct elongates so as to carry the vestibular opening down-

---

* "Anatomische Studien," Bd. iv., "Die Lymphbahnen des inneren Ohres der Wirbelthiere."

wards. The distal enlarged part grows, and, as the tadpole loses its tail, assumes the permanent proportions, becoming at the same time thin-walled and vascular, and the organs from the two sides meet both above and below the brain. Whether actual communication is set up is difficult to determine by means of sections alone.

The growth of cartilage between the expanded end of the organ and the rest of the vestibule does not take place till late, and even then a foramen is left, through which the duct passes from the vestibule into the skull cavity.

The development of this structure shows that the whole of it is part of the internal ear, and is not to be associated morphologically with the lymphatic spaces within the skull. No opening from the expanded portion within the skull is to be found connecting it with other cavities in any way.

f. The Perilymphatic Spaces.—This system of canals has never been described adequately, although it is distinctly developed and very obvious in sections passing through the hinder part of the ear. Hasse and Retzius[*] have both given accounts of parts of these canals; but I believe that they have considered the two to be absolutely continuous with each other, and have described as one what really consists of two distinct parts. The canal which I have taken as the first is carefully described; but the second seems to be taken as forming the anterior part of the first. Its passage into the skull through the foramen rotundum is not described, and the close relation of both to the parts of the cochlea is not noticed. I have investigated the course of these canals in the common frog, the toad, and Dactylethra larva.

In *Rana temporaria* the system is composed of two canals in the hinder part of the ear, closely connected with the cochlea, and communicating with each other within the skull. The walls consist of a single layer of much flattened cells, and backing this layer there are a few scattered cells lying in the perilymph. These walls are distinct though thin at all points in the course of the canals, which therefore do not communicate with any other system.

The first canal to be described, the ductus perilymphaticus, is in close relation with the lagena. The walls of the extremity of this part of the cochlea are very different on the inner and outer sur-

---

[*] "D. Gehörorgan d. Wirbelthiere."

faces, *i.e.*, those looking towards and away from the brain respectively. The outer wall is thickened and bears sensory cells, and it is strongly curved, with the concavity towards the brain. The opposite or inner wall is flat and exceedingly thin, consisting of a layer of very flat pavement cells, and against this flat part abuts the perilymphatic canal. This canal can be traced downwards and backwards as far as a foramen in the inner and hinder part of the floor of the capsule. Through this foramen it passes, and lies between the pharynx and the base of the skull. Here it expands somewhat (saccus perilymphaticus), though it is spacious throughout its course, and from this point it extends backwards some distance, even behind the foramen of the glossopharyngeal nerve. Just in front of this nerve a duct, somewhat narrower than any part yet described, passes up from the perilymphatic space through the foramen along with the nerve, and enters the skull. Its connections within the skull will be described below. The second canal has relations with the pars neglecta of the cochlea, just as the first has with the lagena. It is applied to the flattened wall of this pouch from below; besides this part in immediate connection with the cochlea, there is a spacious continuation of it within the capsule winding past the posterior ampulla, and lying above the hinder part of the utriculus. The remainder of this canal passes through a foramen in the lower, inner, and hinder part of the wall of the capsule (foramen rotundum), and extends back within the brain case till it reaches and communicates with the previously described canal. The part within the skull is of considerable size, occupying the lower angle on each side, and thence it extends up the side wall for some distance, and along the floor nearly to the median line.

The course of these two canals will perhaps be rendered intelligible by Fig. 19. This figure is diagrammatic, and compounded from a large series of sections. By this method the whole course of the system is shown, though it is foreshortened, so that structures are brought into view that could not be cut in one section. Turning now to the other forms in which I have found these structures, I will show that two very similar canals exist in all, though their mutual relations vary.

The toad I have only examined in a young stage just after the tail is lost, but I have no reason to doubt that at this time the adult arrangement obtains; for in frog tadpoles of only 25 mm. in length, these canals have attained to the permanent condition in all essential points. The toad only differs from the frog in the fact that the first canal never leaves the protection of the auditory capsule in its course to the skull. It runs through a special foramen immediately behind the second canal, and in a. position similar to, but somewhat lower than, its point of exit, and so directly into the skull. Within the skull both join as in the frog, and from this common part a large diverticulum runs down through the glossopharyngeal foramen, and occupies a position similar to the corresponding part in the frog. Thus the frog and the toad agree, except in the relative positions of the foramina through which the first canal and the glossopharyngeal nerve pass.

Dactylethra I have cut as a small larva, the legs being present, but as yet neither large nor pentadactylous. In this specimen the perilymphatic spaces are present and connected with the cochlea as in the frog; but they differ from both frog and toad in that they communicate with each other within the capsule, and then pass out by a common duct into the skull. Just where this duct enters the skull a dilatation is present, passing through the foramen of the glossopharyngeal nerve, and lying close above the pharynx as in the frog and toad. I may mention that my Dactylethra specimen seems to show a continuity of these canals of the two sides across the floor of the brain case. I feel sure that this is not so in the other animals examined, and this, together with the imperfect condition of my Dactylethra, makes it very doubtful whether any such connection really exists.

The most interesting point in the development of this system is that in both cases the canals first appear in close connection with the parts of the cochlea with which they are ultimately to be associated. At about 16 mm. the lagena becomes flattened as in the adult, and the beginning of its canal is to be seen as a small sac in the perilymph applied to the lagena as already mentioned. It is formed by mutual separation of the mesoblast cells in the vicinity of the lagena, and the cells so separating constitute the walls of the canal. The second canal appears soon after in a precisely similar manner, and the two grow towards the brain, meet, and unite. The growth of the auditory capsule and skull occurring at this time, separates the inner ear permanently from the portion of the canals within the skull; at the same time foramina are left, through which the communicating ducts pass.

It seems most probable that the canals here described have some connection with the conduction of sound. From their mode of development and their permanent form, it is evident that they are concerned with the cochlea, and it is very possible that the part lying outside the skull above the pharynx receives vibrations, and permits their passage to the auditory labyrinth. If this is the function that the canals perform, they must have been employed in this manner before the columella appeared, and still remained in use after the terrestrial character and structure were acquired.

This suggestion will not explain why the perilymphatic spaces should lie partially within the brain case. I have imagined that this latter peculiarity may be due to their possessing a secondary function of some description other than that already indicated. If this assumption is necessary, their complete meaning is still a problem, notwithstanding the partial solution already offered.

The Eustachian Tube and Tympanic Cavity.—The development of these organs is marked by a peculiar and rapid change of position during the metamorphosis from the tadpole to the frog. The Eustachian tube at the beginning of the change is directed forwards, and its end lies, not in connection with the ear, but under the anterior part of the eye below the palatopterygoid bar. As the tadpole loses its tail, the tube breaks up into short lengths, which move backwards and come to lie in the position which they occupy in the adult. The isolated pieces then join themselves into a connected whole, which by gradual growth becomes the adult organ. Another point to be noted is that in development this tube has almost certainly nothing to do with the hyomandibular cleft. This mode of development in the lowest forms possessing such organs would seem at first sight to prove that the Eustachian tube is not morphologically equivalent to the hyomandibular cleft; but its development in the Anura is so peculiar that it must be regarded as highly modified, and no conclusions should be drawn too positively as to its history. As the Eustachian tube has generally been connected with the hyomandibular cleft in previous accounts of its development, I will here give a few notes on this cleft as found in the frog.

In tadpoles of about 8 mm., the full number of clefts is already recognisable. They consist of five pairs of solid hypoblastic outgrowths, the four posterior pairs having met and blended with the external epiblast, whilst the foremost or hyomandibular cleft ends in the mesoblast some distance from the exterior (see Fig. 6). This stage represents the highest point in the development of the hyomandibular cleft, and after this period it commences to become smaller, never acquiring a lumen for more than a short part of its length, and never opening to the exterior. In tadpoles of a slightly larger size (10 mm.) there is very little trace of the cleft in the form found in the previous stage; but between the hyoid and mandibular arches in the position where the cleft, at the stage just described, joined the œsophagus, a small diverticulum of the gut is to be seen, and this no doubt is the opened-out hyomandibular cleft. This pouch scarcely has any existence of its own, appearing more as a sacculation of the wall of the anterior part of the first branchial cleft than as an independent gill cleft.

It is at this time more ventral than dorsal in position. As the tadpole grows it becomes less and less apparent, until it altogether disappears, or at least can no longer be recognised as bearing any resemblance to a cleft. The latest period at which it may be readily recognised occurs in tadpoles of about 20 mm. It is not until the hyomandibular cleft has ceased to be recognisable that the first trace of the Eustachian tube appears. The Eustachian tube, therefore, is probably not formed from the hyomandibular cleft, but is an altogether new organ. Its development may be divided into two periods, separated from each other by a stage, during which the metamorphosis from the first to the last condition takes place. The first beginning of the tube may be recognised in large tadpoles which have attained approximately their full size, but before the hind legs are conspicuous or toes are apparent. My youngest tadpole in which this organ is present is 25 mm. in length. At this time the Eustachian tube is very poorly developed, and consists of a solid rod of cells running forwards from the dorsal and anterior edge of the first branchial cleft under the palatopterygoid cartilage for a short distance. This rod is very thin and not at all conspicuous; but it may be recognised by its position in relation to the first branchial cleft. Unfortunately, I have cut no sections between this and a considerably later stage; but notwithstanding this, I have no hesitation in identifying the structure just described with the Eustachian tube as it appears in tadpoles just before the time when the fore legs are protruded. When this stage is reached the Eustachian tube has grown forwards, lying close to the palatopterygoid bar, and now it extends to a point just in front of the eye. Though a lumen is not apparent, the cells composing the organ are arranged in the form of a tube with walls more or less definitely one cell in thickness, and at its distal end the tube is somewhat expanded. Figs. 11, 12, and 13 will give an idea of the relations of the structure at this time.

At the commencement of the tadpole's metamorphosis the tube loses its connection with the wall of the pharnyx, and quickly divides into a variable number of short lengths. This state continues until a short time after the frog has lost its tail, and while it lasts the broken fragments move backwards, keeping pace with the shifting of the hyoid arch. When the position of the Eustachian tube that is to be has been reached by its components, they commence to grow towards each other, and at the same time a hollow process of the wall of the pharynx arises and completes the whole length; a cavity is acquired by the walls separating, and the adult form is attained in all important points. This process is shown in Figs. 15 and 16. The irregularity of this change through which the tube passes is very marked. The number of pieces into which it divides differs much in different specimens, some showing only two or three detached lengths, whilst others possess a chain of such lengths consisting of several more links. The two sides of the same specimen often differ considerably. In one specimen the Eustachian tube is complete on one side and has the adult form, whilst the other has only detached fragments (see Fig. 16). No doubt this mode of change from the immature to the adult condition is secondary, and abbreviated from the more natural course of development; the condensation of history being occasioned by the need of a rapid change from aquatic to terrestrial mode of life, accompanied by a rapid alteration of the mandibular and hyoid bars in their position with regard to the rest of the skull. The slightly expanded distal portion of the Eustachian tube described in its second stage is retained to form the future tympanic cavity in all specimens examined, and keeps near the surface, maintaining relations with the future annular cartilage.

The peculiar development of this organ in the lowest forms with such a structure raises a question: Are the hyomandibular cleft and Eustachian tube homologous? If I am correct in stating that there is no connection between the two organs at any period in the frog's history, there is obviously strong reason for doubting the generally accepted doctrine of their homology. But it may be said that the frog's development is modified in this particular point, and that as the two first visceral arches have a peculiar history, in which they probably undergo great larval modification, so they have modified the structure placed between them, *i.e.*, the hyomandibular cleft. The mandibular arch in its early movement, and the hyoid arch by its articulation with the mandibular arch, might cause the hyomandibular cleft to develope in a peculiar way, so that it disappears at one time and again reappears. This is not impossible, for in the second stage described the Eustachian tube occupies a position between the mandibular and hyoid bars at their point of articulation, and thence it runs forwards. Such a course would be peculiar for the hyomandibular cleft; but when the relations of the arches concerned are considered, it will be seen that the early position of the Eustachian tube is not irreconcilable with the view that it is morphologically the hyomandibular cleft. The second point in the development of the tube, that is, its breaking up into pieces during the metamorphosis, may possibly be explained as being due to mechanical causes, originated by the movement of the hyoid and mandibular arches at the time. As far as the considerations advanced go, it is justifiable to hold that the frog's development does not absolulutely disprove that the Eustachian tube and hyomandibular cleft are homologous, although they are probably not connected in development. This view would be in accordance with the doctrine generally held, although a not unimportant section of investigators have considered that actual embryology does not bear out the doctrine usually taught. At the same time it should be remembered that the evidence offered by the frog—and such evidence should have great weight—tends to show that the two organs have no connection whatever with each other.

The Skeleton of the Ear.—In this section is included the development of the annular cartilage, stapes, and columella. Of these the columella is the only one that is of great general interest, from its supposed homology, on the one hand, with the hyomandibular cartilage of fishes, and on the other with the auditory ossicles of higher Vertebrates. Beginning with the annular cartilage, I may preface my remarks by stating that I have obtained traces of it long before the stage is reached which Professor Parker* takes as his earliest. At its first origin it is far in front of the position in which Professor Parker describes it at its first appearance, and it is therefore removed from the columella, and quite distinct from it.

Before the fore legs have appeared, and the tadpole has begun to resemble the frog, the Eustachian tube is directed forwards, and its expanded end, which becomes the tympanic cavity, lies close to the palatopterygoid bar, in advance of the eye (see Figs. 12 and 13). At this time the tympanic cavity is surrounded by a mass of cells, which appear closely packed in contrast with the surrounding loose mesoblastic tissue, and this is especially marked towards the ventral surface of the tube. This mass is closely applied to the palatopterygoid cartilage; but when, during the metamorphosis of the tadpole, the tympanic cavity moves backwards, it carries with it the surrounding mass, which becomes denser; and some time after the metamorphosis has been completed a small cartilaginous bar is formed in it below the ventral edge of the tympanic membrane, and gradually extends until a complete ring grows around the tympanic cavity. This ring is the annular tympanic cartilage. During its growth I can trace no connection between it and the columella close enough to warrant the statement that the two have a common origin. As the columella grows it does come in contact with the annular cartilage; but at first the separation of the two is complete, and even when contact takes place the relation is only one of close juxtaposition.

From the above account, it is clear that the constituent cells of the annular cartilage appear in connection with the anterior part of the mandibular arch, and are perhaps derived from it; although it is possible that they may be merely concentrated from the scattered cells surrounding the distal end of the Eustachian tube. The point at which the cartilage originates is in front of and below the eye, and the part of the mandibular arch to which it is applied is the palatopterygoid bar.

---

* "Structure and Development of Skull in Common Frog," "Phil. Trans.," 1871.

The remaining two skeletal elements are closely connected in the adult, though not actually fused. The stapes forms a concavo-convex cartilage, stopping the fenestra ovalis; whilst the columella is a rod inserted in front of the stapes, between it and the auditory capsule, and expanded distally against the tympanic membrane (see Figs. 17 and 18). Professor Parker calls the proximal end of the columella interstapedial and the remainder the mesostapedial.

The stapes may be dismissed with very few words. It is formed as a chondrification in the capsular membrane closing the fenestra ovalis, at a period when the remainder of the capsule is well developed, and not long before the tadpole begins to assume the frog's form. No more need be said of it here, as Professor Parker has described this cartilage in his account of the frog's skull.

The third and last of the skeletal elements mentioned has now to be considered. As far as I can make out, the columella is embryologically to be considered, not as a part of either the hyoid or mandibular arches, but as similar to the stapes, in that it originates in the membrane closing the fenestra ovalis. In this my results agree with those published in Professor Cope's[*] recent paper, as they also do concerning the separate origin of the columella and annular cartilage.

I have found the columella in the same specimen which showed the origin of the annular cartilage. Just in front of the stapes, and in the membrane connecting it with the remainder of the capsule, a small isolated piece of cartilage is present. This ultimately becomes the inner end of the columella, called by Professor Parker the interstapedial element. From this cartilage a thin rod of cells extends forwards and outwards close to the inner side of the thymus, till it ends just in front of the last-mentioned organ. This rod will ultimately be Professor Parker's mesostapedial cartilage, and by later growth it will expand against the tympanic membrane at its distal end. At the period under consideration it is very thin and not easy to trace, but notwithstanding this I feel sure that it has no connection with any skeletal element except the interstapedial. Under these circumstances it is fair to assume that it either grows outwards from the interstapedial cartilage, or else, with less probability, that it is formed

---

[*] "Hyoid and Otic Elements of the Skeleton in the Batrachia," "Journal of Morphology," vol. ii., No. 2, November, 1888.

in situ by the scattered cells in the vicinity aggregating into the form of a bar. The position of the columella at this period is diagrammatically represented in Fig. 14. It will be seen that it is not at present in connection with the tympanic cavity or annular cartilage, which are indeed far away, nor does it reach the outer surface of the head by some distance. It is also directed much more directly forwards than it is in the adult.

Until metamorphosis takes place, this state of things remains practically unaltered except by slight growth on the part of the organs involved; but as the frog leaves the water, the columella rotates with its base as pivot, so that it comes to point more outwards. At the same time the mandibular and hyoid arches move back, as do the tympanic cavity and annular cartilage. This movement brings the end of the columella into close relation with the ventral surface of the quadrate cartilage near its point of articulation with the capsule, and an attachment takes place between the two by means of a thin connecting bar, the suprastapedial. During this period the columella grows rapidly in length and increases in thickness throughout its whole extent. This causes the distal end to extend close over the tympanic cavity and to come in contact with the annular cartilage very shortly after the frog has completely lost its tail. At the same time chondrification, which has hitherto only appeared at the proximal end, spreads along the whole length (see Figs. 15 and 16). The outer end of the columella now expands against the tympanic membrane, and the cavity below, by growing upwards, causes the cartilage to appear to project through the tympanic cavity on to the membrane. Thus we have the permanent conditions established.

It must be noted that the connecting bar between the mesostapedial and the quadrate cartilage chondrifies and persists, being present even in the adult.

The most interesting point connected with this section applies to the homology of the columella. Firstly, it is not part of the hyoid arch developmentally; nor is it part of the mandibular; for at its first appearance it is connected with neither. The connection that does take place occurs secondarily, and is not a mere temporary affair, as is shown by its size and permanence. It is not to be considered as an embryological, but as an adult connection. The equivalence of the columella with the hyomandibular bar, so strenuously maintained by Professor Parker, receives no support from these researches; but it seems probable that the columella is not only physiologically but morphologically to be associated with the ear capsule, and with it alone.

EXPLANATION OF PLATES XI. AND XII.

Illustrating Mr. Francis Villy's paper on "The Development of the Ear and Accessory Organs in the Common Frog."

Alphabetical List of Reference Letters.

A. C. Annular cartilage. *A. C. H.* Anterior cornu of the hyoid. *Am. A.* Ampulla of the anterior vertical canal. *An. C.* Anterior vertical canal. *Au. C.* Auditory capsule. *Au. E.* Auditory epithelium. *Au. G.* Auditory ganglion. *Au. I.* Auditory invagination. *Au. N.* Auditory nerve. *Au. V.* Auditory vesicle. *B.* Basi-hyal cartilage. *B. C.* Body-cavity. *Br.* Brain. *Br. A.* 1, 2, 3, 4. Branchial arches. *Br. C.* 1, 2, 3, 4. Branchial clefts. *C.* Cerebellum. *Cb.* Cerebral hemispheres. *Cd.* Choroid plexus. *Ch.* Cochlea. *Ch.* 1. Lagena. *Ch.* 2. Pars neglecta. *Ch.* 3. Pars basilaris. *Cl.* Columella. *Cr.* Cornea. *D. A.* Dorsal aorta. *E.* Eye. *Ep. A.* Epithelium of the two anterior ampullæ. *Ep. C.* Epicoracoid. *Eu.* Eustachian tube. *F.* Fenestra ovalis. *G. G.* Gasserian ganglion. *G. N.* Glossopharyngeal nerve. *Gu.* Gullet. *Hb.* Hind-brain. *Ho. A.* Ampulla of the horizontal canal. *Ho. C.* Horizontal canal. *Hp.* Hypoblast. *Hu.* Humerus. *Hy. A.* Hyoid arch. *Hy. M.* Hyomandibular cleft. *In.* Interstapedial cartilage. *Ir.* Iris. *K.* Kidney. *L.* Lens. *Md.* Mandibular arch. *Mk.* Meckel's cartilage. *Mo.* Mouth. *Ms.* Mesostapedial cartilage. *Ne.* Nervous layer of epiblast. *N. G.* Neural groove. *No.* Nose. *Nt.* Notochord. *Ol. N.* Olfactory nerve. *Op. l.* Optic lobes. *Op. N.* Optic nerve. *Ot. E.* Outer layer of epiblast. *Pa. pt.* Palatopterygoid bar. *Pc.* Precoracoid. *Pe.* 1. Perilymphatic canal (connected with the lagena). *Pe.* 2. Perilymphatic canal (connected with the teg. vasc.). *P. V. A.*

Ampulla of the posterior vertical canal. *P. V. C.* Posterior vertical canal. *Q.* Quadrate cartilage. *R.* Recessus labyrinthi. *S.* Scapula. *Sac.* Sacculus. *S. A.* Septum of anterior vertical canal, outer fold. *S. H.* 1. Septum of horizontal canal, upper fold. *S. H.* 2. Septum of horizontal canal, lower fold. *S. P.* 1. Septum of posterior vertical canal, outer fold. *S. P.* 2. Septum of posterior vertical canal, inner fold. *St.* Stapes. *Stn.* Sternum. *Su.* Suprastapedial cartilage. *Su. s.* Suprascapula. *T.* Tail-muscles. *Th.* Thymus. *Ty.* Tympanic cavity. *U.* Utriculus. *V.* Vagus. *iii.* Third nerve.

These figures are all more or less diagrammatic in different ways, but in all the general outlines are as nearly correct as possible, for they are all taken from camera drawings. The shading is diagrammatic, and no attempt has been made to show cells, except in a rough way in the younger stages of the auditory epithelium. In other organs, such as the young brain, any marking that may suggest cellular structure is only used for the sake of distinction. To avoid confusion, the mesoblastic tissues, except those immediately concerned, are for the most part coloured uniformly. The perilymph and skull cavities are finely dotted; whilst cartilage is denoted by coarser dots.* Fig. 6 and Figs. 13–19 are formed by compounding a number of sections so as to represent a single ideal and actually impossible one. In Figs. 15 and 16 the columella is the only part to which this description applies.

Fig. 1. Transverse section through the auditory region of an embryo frog, just before the coalescence of the neural folds. The auditory nerve and epithelium are already recognisable, although invagination has not yet begun.

Fig. 2. Similar section of a slightly older embryo, with the neural groove closed. On the left hand side the section passes through the centre of the invagination, and cuts the whole length of the auditory nerve. Invagination is more advanced at the dorsal edge than it is at the ventral. On the right the section passes through the anterior part of the auditory involution.

---

* The endolymph and contents of the perilymphatic canals are left white.

Fig. 3. Similar section of an older embryo. The auditory invagination is cut through the centre on the left, and is more advanced than it is in Fig. 2. The nerve is shown.

Fig. 4. Similar section of an older embryo shortly before hatching, i.e., about 4 mm. in length. Invagination is now completed, and the vesicle is pyriform.

Fig. 5. Slightly oblique transverse section of a tadpole of 8 mm. On the left the section passes through the centre of the auditory vesicle, and the recessus labyrinthi is shown. On the right the section passes through the anterior part of the vesicle, showing the auditory ganglion and the epithelium common to the two anterior ampullæ separating from the rest.

Fig. 6. Diagrammatic horizontal section of a tadpole of 8 mm., to show the gill clefts and arches. The fore- and mid-brain are cut, as well as the nose and kidney. The five pairs of clefts and six pairs of arches are represented. The hyomandibular cleft does not reach the skin.

Fig. 7. Oblique transverse section of a tadpole of 12 mm., showing the beginnings of the septa which cut off the semicircular canals. On the left the recessus is cut, along with the outer and upper septa belonging to the anterior vertical and horizontal canals respectively. On the right the anterior part only of the recessus is cut, and the outer ridge of the septum of the anterior vertical canal, together with the lower part of the septum of the horizontal canal: ($\times$ 25).

Fig. 8. Section through a somewhat older tadpole than the last. On the right the horizontal canal is cut, its septum being completed. On the left the folds forming the septum of the posterior vertical canal are shown; they have not yet met. Part of the beginning of the septum between utriculus and sacculus is represented, as is the small pouch representing the lagena: ($\times$ 25).

Fig. 9. Transverse section through the centre of the ear in a tadpole of 25 mm. The full extent of the recessus at this stage is shown: ($\times$ 15)

Fig. 10. Oblique transverse section through the head of a large tadpole, just before the anterior legs are protruded. On the left the first branchial cleft is shown, as well as the opening of the Eustachian tube into the œsophagus. On the right hand the centre of the eye is cut, and a section of the Eustachian tube is shown considerably in front of its point of origin from the throat: (× 10).

Fig. 11. Section through the tympanic cavity and annular cartilage in the same specimen. On the left it passes entirely in front of the tympanic cavity, cutting the point of union of the palatopterygoid bar with the trabeculæ. The section is completely in front of the eye: (× 12).

Fig. 12. Diagrammatic sagittal section of a tailed frog on the left side, showing the course of the Eustachian tube and the position of the annular cartilage. It passes through the ear and eye and the angle of the mouth. The greater part of the length of the quadrate and palatopterygoid cartilage is shown with the Meckelian cartilage below it, as well as the outer portions of the hyoid and first three branchial arches: (× 15).

Fig. 13. Oblique section through the ear of the tadpole from which Figs. 10 and 11 are taken. The columella is shown at an early stage cut at two different levels. On the right the interstapedial, and on the left the mesostapedial is shown. The thymus lies a short distance to the outer side of the columella: (× 10).

Fig. 14. Diagrammatic horizontal section of a tailed frog, to show the course of the columella. The eye is cut some distance from its centre, and the anterior nares are included in the section. The relations of the columella to the auditory capsule, stapes, and thymus are shown: (× 15).

Fig. 15. Oblique transverse section of a young frog's head, just after the metamorphosis. The left side is more anterior than the right. On the right the whole course of the columella is shown; it is as yet chondrified at its base alone. Observe the fragmentary and asymmetrical condition of the Eustachian tube: (× 15).

Fig. 16. Section similar to the last, but through an older frog. The columella is now chondrified throughout. Observe that the Eustachian tube is complete on the left, but fragmentary on the right: ($\times$ 15).

Fig. 17. Diagrammatic transverse section of an adult frog's head. The whole course of the recessus is shown, with the connection between the two sides, both above and below the brain. On the right as much of the columella is shown as possible, and the whole course of the Eustachian tube is represented. The anterior parts of the sternum and thymus are cut, and the anterior cornu of the hyoid is shown at different levels on the two sides: ($\times$ 8).

Fig. 18. Diagrammatic horizontal section of an adult frog's head. The length of the recessus is shown on the right, and the columella and stapes are in position. On the left an attempt has been made to show the distribution of the auditory nerve to the various parts of the ear. The nerve bifurcates within the skull, and the anterior branch supplies the sacculus, the utriculus, and the two anterior ampullæ. The posterior branch runs to the three parts of the cochlea and to the posterior ampulla. On the right the two perilymphatic canals are seen. Besides the parts mentioned, the section cuts the centre of the eye and of the cerebral hemispheres, the lower part of the optic lobes, the nose, and the Gasserian ganglion: ($\times$ 8).

Fig. 19. Diagrammatic transverse section through the posterior part of an adult frog's ear, to show the course of the perilymphatic canals. In this figure the whole of the pectoral arch is shown, along with the head of the humerus. The posterior part of the stapes appears, as do the posterior vertical and the horizontal canal (cut twice), the posterior ampulla, and the two parts of the cochlea concerned. The brain is cut through the anterior part of the medulla oblongata, and the choroid plexus is shown: ($\times$ 8).

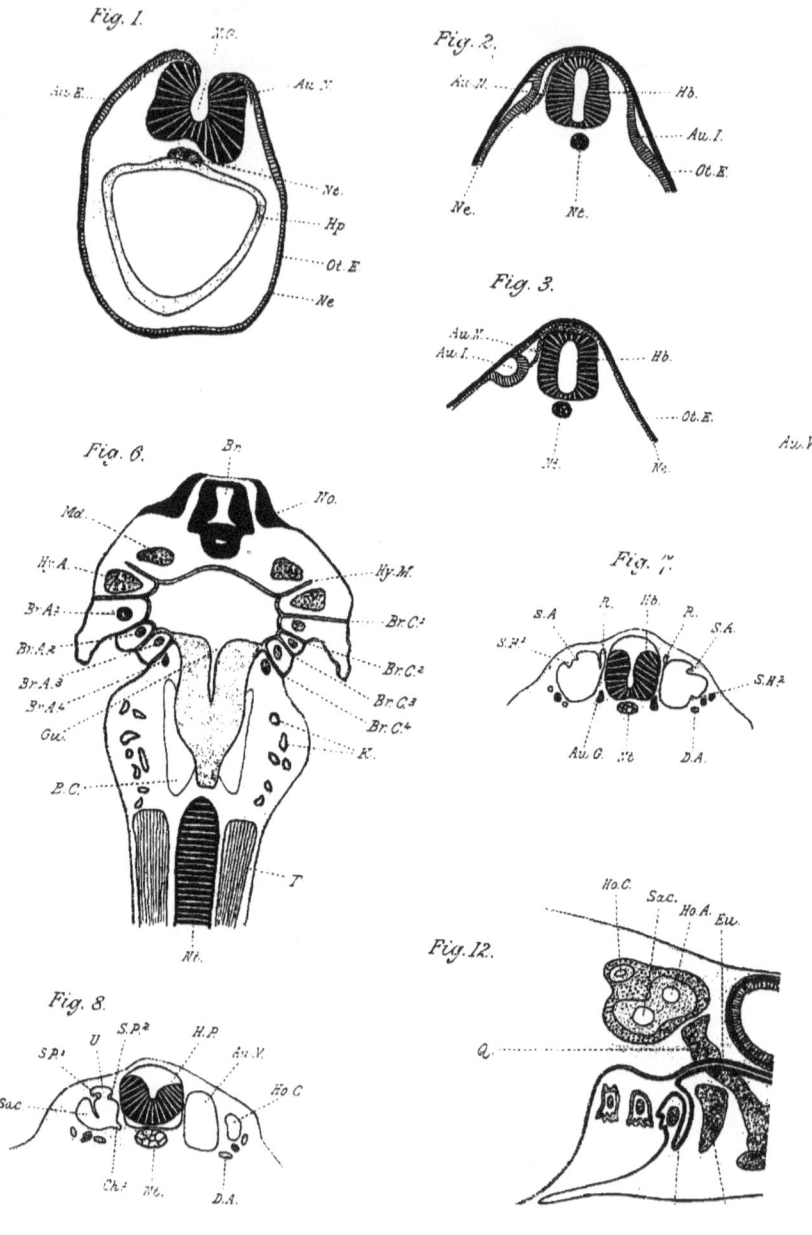

Plate XI

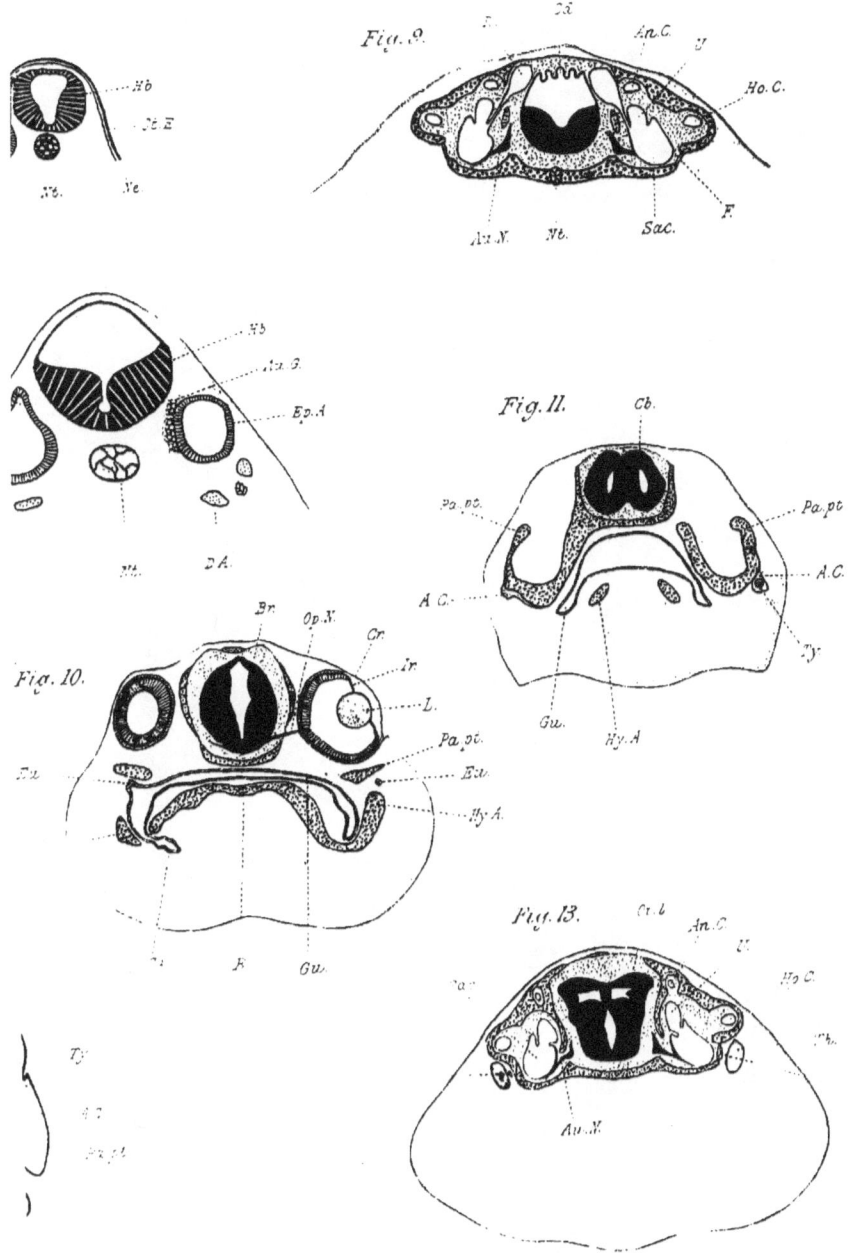

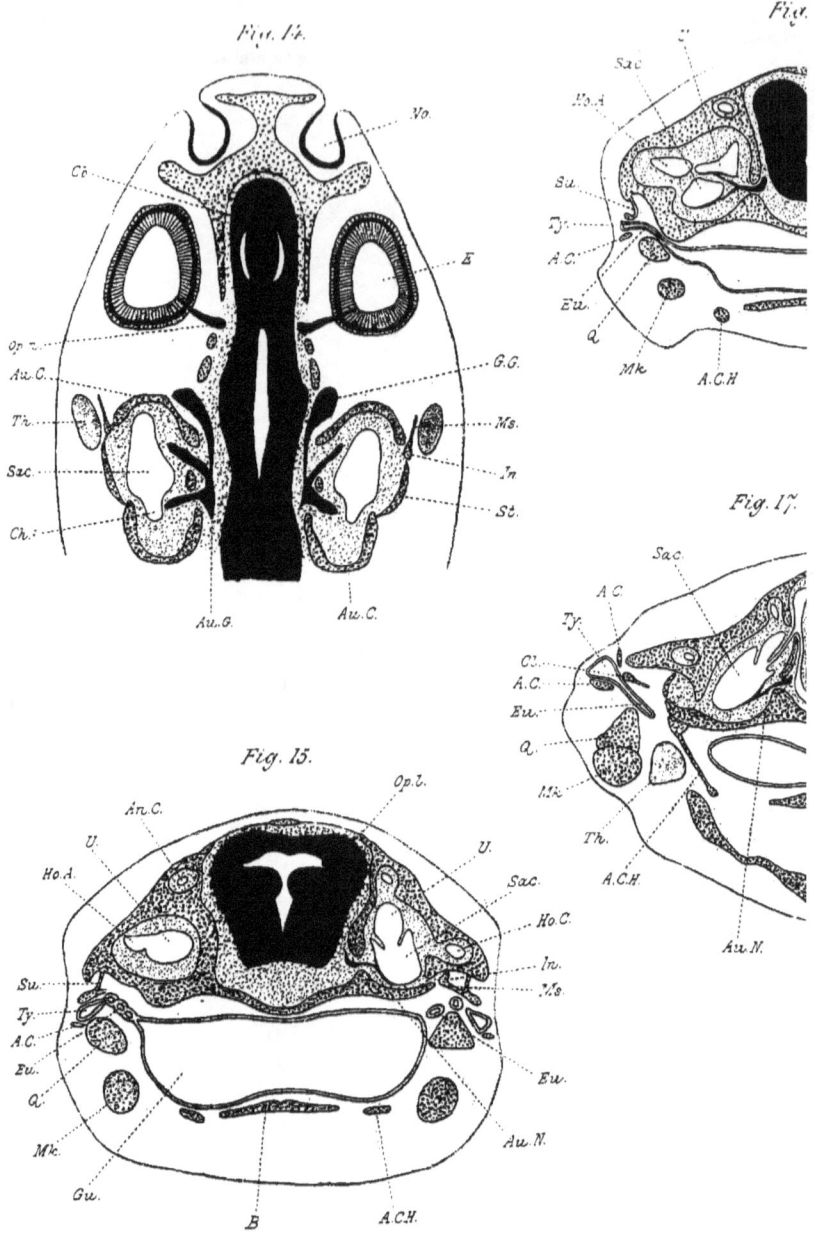

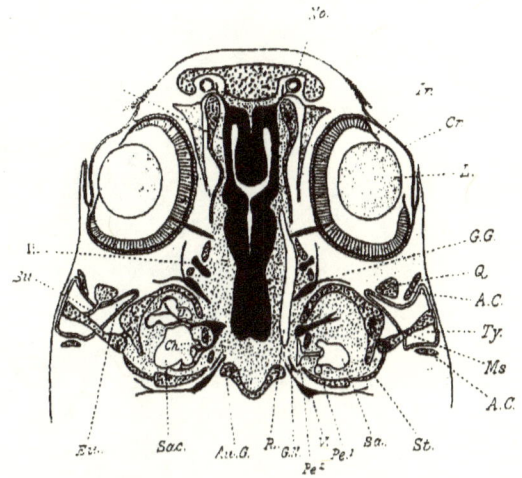

Fig. 18.

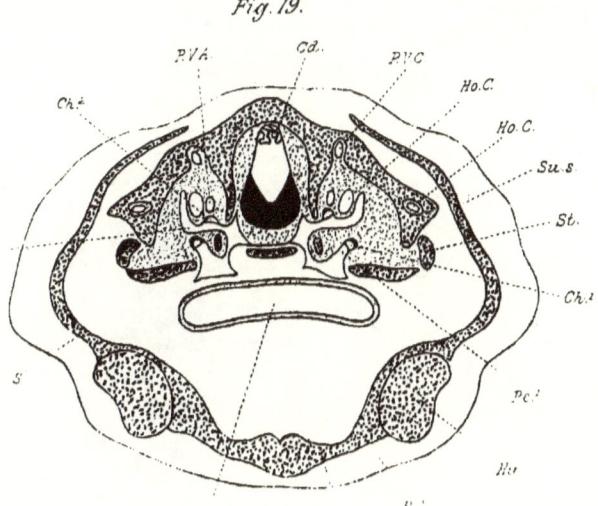

Fig. 19.

THE DEVELOPMENT OF THE BLOOD VESSELS IN THE FROG.

By A. MILNES MARSHALL, M.D., F.R.S., *Beyer Professor of Zoology in the Owens College;* and
EDWARD J. BLES, *of the Owens College.*

[WITH PLATES XIII., XIV., AND XV.]

CONTENTS.

	PAGE
Introduction	186
Historical Account	187
The Early Development of the Heart	190
The Heart and Blood Vessels in $4\frac{1}{2}$ mm. Tadpoles	196
The Heart and Blood Vessels in 5 mm. Tadpoles	208
The Heart and Blood Vessels in $6\frac{1}{2}$ mm. Tadpoles	215
The Heart and Blood Vessels in 9 mm. Tadpoles	222
The Heart and Blood Vessels in 12 mm. Tadpoles	231
The Heart and Blood Vessels in 21-23 mm. Tadpoles	248
The Heart and Blood Vessels during the Metamorphosis	252
The Heart and Blood Vessels in First Year's Frogs	257
Summary	259
Description of the Figures	261

INTRODUCTION.

The development of the blood vessels, and more especially of the aortic arches, in Amphibians is a subject of peculiar interest, owing to the fact that members of this group repeat in their individual history the transition from aquatic to aërial respiration, which there is every reason to regard as ancestral, not merely for themselves, but for all the higher Vertebrates as well.

If we are right in interpreting individual development as representing, in however modified a form, the past ancestral history of the race, then the study of the respiratory change in Amphibians ought to throw an important light on the mode in which the transition first came about. Moreover, the actual change is in Amphibians a gradual one, the gills and lungs being used simultaneously for a considerable period of time, or in many forms throughout life, so that there is abundant opportunity for studying quietly and deliberately the details of the transformation.

In spite, however, of the interest of the problem, and of the ease with which suitable material for investigating it may be obtained, there are many points in actual development that are still very imperfectly understood, and it was in the hope of clearing up some of the doubtful or disputed matters that the investigations were undertaken, the results of which are recorded in the present paper.

For material we have employed almost exclusively embryos and tadpoles of the common frog, *Rana temporaria;* including all the more important stages from the first appearance of the blood vessels to what is practically the adult condition. With regard to methods of investigation, the examination of living specimens has been of considerable use; but all our more important results were obtained from dissections and from the study of sections. Of the latter a very large series was prepared, specimens of all the stages being cut in the three principal planes,—transverse, horizontal, and sagittal. For hardening embryos, we have found Kleinenberg's picro-sulphuric acid, and corrosive sublimate the most useful reagents; for staining we have chiefly employed borax carmine, but in some cases have obtained very excellent results with aniline blue-black, after hardening with corrosive sublimate.

The figures illustrating this paper are in almost all cases diagrammatic, inasmuch as each has been constructed by combining a number of camera drawings of separate sections, so as to include parts situated at different levels. Considerable pains have been taken in each case to represent the shapes and positions of the parts shown as correctly as possible.

HISTORICAL ACCOUNT.

We do not propose to attempt an exhaustive consideration of the literature bearing on our subject. A few of the more important memoirs will alone be noticed here, and reference will be made from time to time in the text to the statements and conclusions of other investigators.

Of the earlier accounts of the development of the blood vessels in Anura, by far the most important is that of Rusconi,* who, relying entirely on dissections and on examination of living specimens, succeeded in determining a number of difficult points, and whose descriptions have afforded the basis for the accounts given by all writers until very recent times. Rusconi distinguished between the afferent and efferent branchial vessels in each arch, naming them "artères transitoires" and "artères permanentes" respectively, and describing carefully and correctly their positions in the arches and in the gill loops, and their relations to the vessels of the adult frog. He described and named the "filters" on the gill arches, and noted the comparatively late appearance of the fourth aortic arch, and the development from it of the pulmonary artery. He was wrong, however, in describing the afferent and efferent branchial vessels of each arch as remaining in direct communication with each other throughout the whole larval period.

Ecker,† in his "Icones Physiologicæ," gives a couple of figures, but no detailed description of the blood vessels of the tadpole.

---

* Rusconi, M., "Développement de la Grenouille commune depuis le moment de sa naissance jusque à son état parfait," Milan, 1826, pp. 47-55.

† Ecker, "Icones Physiologicæ," Leipzig, 1851-59.

Huxley, in his "Anatomy of Vertebrates," gives the following account of the changes undergone by the branchial vessels during development.* " When the internal gills of the Batrachia appear, each aortic arch which belongs to a branchial arch splits into two trunks,—one which remains directly connected with the cardiac aorta, and another which opens into the dorsal aorta. The vessels of the branchial filaments constitute loops between these afferent and efferent trunks, which always remain united by anastomoses. When branchial respiration ceases, and the branchial processes and their vessels disappear, the anastomoses dilate; the direct communication between the afferent and efferent trunks of the second pair of internal branchiæ is re-established; and they become the permanent arches of the aorta. The anterior branchiæ are replaced by the carotid glands, and their afferent vessel is the carotid passage of the adult. The afferent and efferent trunks of the third pair of branchiæ are converted into the stem of the cutaneous artery, and the afferent trunk of the fourth pair of branchiæ into that of the pulmonary artery." This account has been very generally accepted in this country, but is in many respects at variance with the conclusions at which we have arrived.

Goette† gives a careful and detailed description of the development of the heart, dealing more particularly with the early stages of its formation. He also describes at some length the development of the venous system. His account of the arteries, and especially of the branchial vessels, is, however, very unsatisfactory; and with regard to the important changes undergone by these vessels during the transition from branchial to pulmonary respiration, he gives no observations of his own, but merely states that they occur in the manner described by Rusconi.

Boas‡ in a paper on the "Conus Arteriosus and Aortic Arches of Amphibia," gives some important details concerning the development

---

* Huxley, "A Manual of the Anatomy of Vertebrated Animals," London, 1871, p. 192.

† Goette, A., "Die Entwickelungsgeschichte der Unke: Bombinator igneus." Leipzig, 1875., pp. 745-788.

‡ Boas, "Ueber den Conus Arteriosus und die Aortenbogen der Amphibien," "Morphologisches Jahrbuch," vii., 1881, pp. 488-572.

of the branchial vessels. His observations, which were based on dissection, aided by injection of the vessels, deal only with the later stages of larval development; and his principal results are that both the pulmonary and the cutaneous arteries arise from the fourth aortic arch, the third arch being present in the earlier stages, but atrophying about the time of the metamorphosis.

In a subsequent paper Boas* notes that the carotid gland is not, as has been very commonly assumed, a persistent and altered portion of a gill, but is a structure arising altogether independently of the gill vessels.

By far the most important paper that has yet appeared on the development of the Amphibian blood vessels, is one by Maurer,† published two years ago. This is a very thorough piece of work, of great value, and deals with the formation of the blood vessels from their earliest appearance. Maurer's results were obtained chiefly from the examination of continuous series of sections, and his paper is illustrated by a number of excellent and exact figures. We had made considerable progress with our work before becoming acquainted with Maurer's description; but from the time of our doing so, we have found his paper of the greatest service. We shall have occasion to refer frequently to Maurer's work, but it will be convenient to give here a brief summary of his results.

Maurer describes in each branchial arch two vessels, which he names primary and secondary aortic arches, and which correspond respectively to the artères permanentes and artères transitoires of Rusconi, and to the efferent and afferent vessels of our own account. The primary aortic arches arise independently of the heart, and are at first more or less lacunar. They acquire connections with the heart and with the dorsal aorta before the appearance of the secondary aortic arches, and so afford for a time a direct passage from the heart to the dorsal aorta.

The secondary aortic arches arise as direct outgrowths from the divisions of the truncus arteriosus, and lie in the ventral parts of

---

* Boas, "Beiträge zur Angiologie der Amphibien," "Morphologisches Jahrbuch," viii., 1882, pp. 169-187.

† Maurer, "Die Kiemen und ihre Gefässe bei Anuren und Urodelen Amphibien," "Morphologisches Jahrbuch," xiv., 1888, pp. 175-222.

the branchial arches, immediately behind the corresponding primary aortic arches. From the secondary arches outgrowths arise which open into the primary arches, and which form the capillary loops of the gills, both external and internal. The primary arches lose their connection with the truncus arteriosus, and the course of the blood in each branchial arch is now from the heart to the secondary arch or afferent branchial vessel, then through the capillary loops of the gill into the primary arch or efferent branchial vessel, and so on to the dorsal aorta.

At the time of the metamorphosis the primary arches re-acquire their connection with the truncus arteriosus, and so afford a direct passage from the heart to the dorsal aorta, by following which the blood can escape the gill capillaries altogether. The gills and the secondary arches atrophy, while the primary arches remain as the permanent aortic arches of the adult.

Maurer agrees with Boas in deriving both the pulmonary and the cutaneous arteries from the fourth arch, the third arch atrophying at the time of the metamorphosis.

Our own results agree very closely with Maurer's, which they confirm in most of the essential points. Some differences occur, which will be noticed in due course.

I. THE EARLY DEVELOPMENT OF THE HEART.

The heart in its early stages of development lies very far forward on the under surface of the head, below the floor of the pharynx, above and slightly behind the sucker, and immediately in front of the commencing liver.

Its position and relations in an embryo of $3\frac{1}{2}$ mm. length are shown in sagittal section in Plate XIII., Fig. 2, and in transverse section at a slightly later stage in Fig. 3.

In Fig. 3 the mesoblast at the sides of the pharynx is unsplit, but in the floor of the pharynx, on the ventral surface of the embryo, it is divided into somatic and splanchnic layers, M O and M P, separated from each other by a considerable space, C A, which will become later on the pericardial cavity. Of the two layers of mesoblast, the outer or somatic layer keeps close to the epiblast; the inner or splanchnic layer is in close relation with the hypoblast at the sides of the pharynx, but beneath the middle part of its floor is separated from it by a considerable interval, so that beneath the floor of the hinder part of the pharynx (*cf.* Figs. 2 and 3) there is a space, somewhat quadrangular in transverse section, bounded above by hypoblast, and below and at the sides by the splanchnic layer of the mesoblast.

Within this space lies a tube with thin walls, H, which is free along the greater part of its length, but attached in front to the hypoblast of the floor of the pharynx (Fig. 2, H), and continuous behind with the anterior surface of the liver rudiment. This tube forms the endothelial lining of the heart. It consists of a single layer of cells, somewhat variable in shape and size, but the majority of which are slightly flattened in shape, and in contact with their neighbours at their edges. At the hinder end of the endothelial tube the cells are more spherical and more richly laden with yolk granules than those of the anterior part; and a few of these spherical cells are commonly seen lying within the cavity of the tube (Figs. 2 and 3, H). The tube, as seen in Fig. 2, H, is closed in front by a wall of its own, distinct from the hypoblast of the floor of the pharynx; at the hinder end the tube has no endothelial wall, but is closed by the anterior surface of the liver (Fig. 2).

The muscular wall of the heart is formed from the splanchnic mesoblast (Figs. 2 and 3, H B), which, at the present stage, consists of a single layer of short columnar cells surrounding the endothelial tube, H, below and at the sides. Later on, this mesoblast grows in from either side towards the middle line between the floor of the pharynx and the endothelial tube. In this way the mesoblastic investment of the heart, from which the muscles are formed, is completed, as is also the mesoblastic wall of the pharynx. For a time these two structures remain connected by a median vertical mesocardial fold, formed by the meeting of the mesoblastic ingrowths from the two sides; later on, this mesocardial fold is absorbed.

So far the account we have given of the formation of the heart is in harmony with the descriptions of all investigators. Concerning the earlier stages of development, there is, however, great difference of opinion, more particularly with reference to the origin of the cells which form the endothelial lining of the heart.

Oellacher* commences his account of the development of the heart in Bufo at a stage a little earlier than that shown in Plate XIII., Fig. 3. He describes and figures the whole of the cavity below the pharynx, between hypoblast and mesoblast, as filled up by a solid mass of cells, whose origin he was unable to determine, and from some of which the endothelium of the heart is derived. His figures suggest that what he describes as a solid mass of cells is really, in great part at least, a coagulum; and his observations did not begin at a sufficiently early stage to enable him to determine the origin of the endothelium.

Goette† describes the endothelial cells in Bombinator as forming at first a loose discontinuous layer of cells derived from the hypoblast of the pharyngeal wall, but possibly including also cells derived from the mesoblast of the splanchnopleure.

Balfour‡ says, concerning the cardiac endothelium of Elasmobranchs: "The origin of this lining layer I could not certainly determine, but its connection with the splanchnic mesoblast suggests that it is probably a derivative of this." Concerning the Fowl, he says: "I have been able to satisfy myself that the epithelioid lining of the heart is derived from the splanchnic mesoblast. When the cavity of the heart is being formed by the separation of the splanchnic mesoblast from the hypoblast, a layer of the former remains close to the hypoblast, but connected with the main mass of the splanchnic mesoblast by protoplasmic processes. A second layer next becomes split from the splanchnic mesoblast, connected with the first layer by the above-mentioned protoplasmic processes. These two layers form the epithelioid lining of the heart; between them is the cavity of the heart, which soon loses the protoplasmic trabeculæ which at first traverse it."

Rabl§ notices that in Salamandra and Triton a median longitudinal groove is present in the floor of the pharynx, extending back-

---

* Oellacher, "Ueber die erste Entwickelung des Herzens und der Pericardialoder Herzhöhle bei Bufo cinereus:" "Arch. f. Mikr. Anatomie," vii., 1871, pp. 157-165.

† Goette, "Entwickelungsgeschichte der Unke," 1875.

‡ Balfour, "Elasmobranch Fishes," 1878, p. 230, and "Comparative Embryology," 1881, vol. ii., p. 521.

§ Rabl, "Ueber die Bildung des Herzens der Amphibien:" "Morphologisches Jahrbuch," xii., 1887, pp. 252-274.

wards a short distance from the rudiment of the thyroid body. Beneath this groove the hypoblast of the pharynx and the external epiblast are in close contact along the median line, though right and left of this they are separated by mesoblast which has already split into somatic and splanchnic layers. It is immediately behind this region that the heart is formed, and Rabl, who maintains a hypoblastic origin for the cardiac endothelial cells, is inclined to connect these genetically with the down-growing ridge of hypoblast immediately in front.

Rückert* describes the endothelial cells of the heart and great vessels as arising, in Selachians, partly from hypoblast and partly from mesoblast, the first source of origin preponderating in some regions, the second in others.

The most recent investigator of this point is Schwink,† who has studied the early development of the heart in both Anura and Urodela, the two groups agreeing in all essential respects. He notes that the cardiac endothelial cells must be derived either, (1) from the splanchnic mesoblast; or (2) from the hypoblast of the floor of the pharynx; or (3) from the hypoblast further back, in the region of the liver. In attempting to decide between these, he pays special attention to the forms and characters of the cells, and to the evidence of active division afforded by karyokinetic nuclear figures. The endothelial cells themselves have no constant form: some are spindle-shaped, with two or more processes; others are spherical and of various sizes; and all contain numerous yolk granules. On the other hand, the hypoblast of the ventral wall of the pharynx, and the splanchnic mesoblast alike consist of single layers of short columnar cells of definite shape and size, and closely packed side by side. Nuclear figures are not numerous in the cells of these layers, but those that do occur indicate division in the planes of the layers, and not at right angles to these planes, as must be the case if the endothelial cells are derived from one or other of these layers. Schwink accordingly concludes that neither

---

* Rückert, "Ueber die Entstehung der Endothelialen Anlagen des Herzens und der ersten Gefässstämme bei Selachier Embryonen;" "Biologisches Centralblatt," viii., 1889, pp. 385-399 and 417-430.

† Schwink, "Ueber die Entwickelung des Herzendothels der Amphibien," "Anatomischer Anzeiger," V., 5 April, 1890, pp. 207-213.

the hypoblast of the pharyngeal wall nor the splanchnic mesoblast give origin to the cardiac endothelial cells.

On the other hand, he finds that at the posterior end of the heart the conditions are different; here the heart lies close against the commencing liver, the walls of which consist at this stage of yolk cells of varying shape and size. The posterior wall of the heart is formed, as already noticed, by the liver itself (*cf.* Plate XIII., Fig. 2). At this part karyokinetic figures are numerous, indicating rapid cell division, and it is from these cells that Schwink believes the endothelial cells of the heart are formed. He therefore assigns to these cells a hypoblastic origin; agreeing in this respect with Goette and Rabl, though differing from both these writers in regard to the particular region from which they are derived.

Our own investigations are not altogether conclusive on the point. We find that in tadpoles measuring from 3 mm. to $3\frac{1}{4}$ mm. in length, while the mesoblast in the trunk forms a continuous and well-defined layer surrounding the body, except in the mid-dorsal part, the condition is different in the pharyngeal region. Plate XIII., Fig. 1, represents a transverse section of a $3\frac{1}{4}$ mm. tadpole, the section passing through the auditory sacs above and the developing suckers below. The mesoblast consists dorsally of scattered cells, mostly stellate in form; at the sides of the pharynx the cells are more spherical in shape and more closely compacted; they are also more or less definitely arranged in outer and inner layers. Towards the ventral surface these layers—somatic, M O, and splanchnic, M P—become more clearly defined, and are separated by a very distinct space, the future pericardial cavity, C A. The two sheets of mesoblast, of the right and left sides of the body, approach very close to each other in the mid-ventral line, but do not quite meet. The two layers—somatic and splanchnic—into which the mesoblast is split on each side differ very markedly. The outer or somatic layer, M O, lies close to the epiblast, and consists of a single layer of flattened cells. The inner or splanchnic layer, which consists of a single stratum of rather large columnar cells, lies close to the hypoblast opposite to the sides of the pharynx; but on reaching the ventral surface of the pharynx it bends away from the hypoblast rather sharply, and then again turns inwards towards the median plane. The space that is thus left between the hypoblast and the splanchnic mesoblast is the cavity of the future heart. In this space lie a number of cells, H, variable in size, but mostly spherical in shape; they have large nuclei, and are richly studded with yolk granules.

These are clearly the cells which are destined at a slightly later stage to form the cardiac endothelium. Concerning their origin, we have always found them to be perfectly distinct from the mesoblast, while in several cases they are in very close relation with the hypoblast. In some instances their connection with the hypoblast is a very intimate one, and we have noticed endocardial cells apparently in the act of separating off from the outer ends of the hypoblast cells. In this region there is, as Rabl has pointed out, a slight median longitudinal groove on the floor of the pharynx, continuous anteriorly with the thyroid depression. This groove causes a slight keel-like projection on the under surface of the pharynx, and it is in connection with this keel that we have found the appearance of budding off of endocardial cells from the hypoblast to be best marked. The section drawn in Fig. 1 is taken through the hinder end of this groove, which is here hardly recognisable.

The hinder end of the heart is from the first continuous with the vitelline veins that form on the surface of the yolk. The endothelial cells of this hinder part are at this stage larger and more spherical than those further forward, and we are inclined to think that in this region the cardiac endothelial cells are formed in the manner described by Schwink.

We have found no evidence, however, of a shifting forward of the endothelial cells, such as must occur if all the endothelial cells arise in the region of the liver. On the other hand, we have, as explained above, obtained evidence that the endothelial cells of the anterior part of the heart arise directly from the hypoblast of the ventral wall of the pharynx.

Concerning the hypoblastic origin of the endothelium in the frog, while fairly assured of its actual occurrence, we think it would be a mistake to attempt to assign to it any great morphological significance. In many respects the development of the frog, especially in the early stages, gives evidence of secondary modification of an extensive kind. It will be sufficient to call to mind the two-layered condition of the epiblast, and the consequent peculiarities in the development of the ear and other organs. The mode of formation of the mesoblast is also in many respects very far removed from the typical Vertebrate plan. We are inclined, therefore, to regard the cardiac endothelium, not as consisting of hypoblast cells, but of mesoblast cells, formed from the hypoblast in a somewhat unusual manner, and at a place where we have shown above there are no other mesoblast cells.

The statement that has been recently made* that the notochord in the frog is of mesoblastic and not of hypoblastic origin, is, we believe, misleading, for the same reason, and due to a failure to appreciate the great modification which the early development of the frog has undergone. We have paid some attention to the point, and find the notochord at its posterior end to be clearly and distinctly hypoblastic, and in its anterior part to be formed from a mass of cells concerning which it is impossible to determine whether they are to be spoken of as hypoblastic or mesoblastic.

II. THE CONDITION OF THE HEART AND BLOOD VESSELS IN 4½ MM. TADPOLES.

In describing the later phases in the development of the heart, and the mode of formation of the principal blood vessels, we propose to take a series of stages selected somewhat arbitrarily, but which we have found to represent in a convenient manner the principal steps in the history. We have defined these stages by the length of the tadpole; but as there is a considerable amount of variability in this respect, we have in each case given a brief account of the external features, and of the leading points of internal structure, which we trust will render the determination of our stages a comparatively easy matter.

Our figures are in almost all cases highly diagrammatic. They have been constructed in order to facilitate as far as possible comparison of the several stages with one another. In each figure many important organs are entirely omitted, but the parts which are shown have been drawn with great care, and their proportions and relative positions are shown accurately.

---

* Schultze, O., "Die Entwickelung der Keimblätter und der Chorda dorsalis von Rana fusca," "Zeitschrift für wissenschaftliche Zoologie," Bd. 47, 1888, pp. 325-352.

Our first stage is taken shortly before the tadpole works its way out of the gelatinous mass of the spawn and becomes free. At this stage our tadpoles have an average length of $4\frac{1}{2}$ mm.

1. External Characters.

The head is large and prominent, and separated by a very distinct neck constriction from the body. The body is greatly distended ventrally by the yolk mass. The tail is well formed, but short, not exceeding a fourth of the total length of the animal.

The sucker is well-developed, and forms a prominent object on the ventral surface of the head. The olfactory organs form a pair of shallow circular depressions, with slightly raised margins, at the anterior end of the head. The eyes are already present, but are only visible on the surface as a pair of slightly raised longitudinal ridges, immediately behind the olfactory pits, and continuous with their raised margins. Below the olfactory pits, in the median line, is the stomatodæum, which as yet is a small and very shallow depression. Two pairs of external gills are present as backwardly directed processes from the first and second branchial arches; they are somewhat conical in shape, with rounded or very slightly notched hinder borders. The gill of the first arch overlaps that of the second, and is placed rather more ventrally than the latter.

In front of the first gill the hyoid arch may be recognised as a slight vertical ridge, and behind the second gill the third branchial arch is present, though inconspicuous, owing partly to its lying in the constriction between head and body, and partly to its being overlapped by the second gill.

The myotomes are clearly visible along the whole length of the body and tail; they are most distinct a short distance behind the head, and are > shaped, with the angles directed forwards as in the adult Amphioxus.

2. Internal Anatomy.

The pharynx is of considerable size, being both wide transversely and deep dorso-ventrally (*cf.* Plate XIV., Fig. 5). In horizontal section it is somewhat diamond-shaped, its anterior angle being prolonged forwards beneath the infundibulum as a hollow, laterally compressed diverticulum which is in close relation with the epiblast of the stomatodæal invagination.

There are five gill pouches on each side of the pharynx. These pouches are double folds of hypoblast, forming vertical partitions projecting outwards from the pharynx to the surface epiblast. The cavity of the pharynx extends a very short distance outwards between the two layers of each fold, and along almost their entire extent the two layers are in close contact with each other. The outer ends of the pouches reach the epiblast and fuse with its inner or nervous layer.

As seen in horizontal section, the gill pouches radiate outwards from the centre of the pharynx. The most anterior pair are the hyomandibular pouches or clefts; and then in succession come the first, second, third, and fourth branchial clefts, the hindmost or ourth branchial cleft being always small, and in some specimens indistinct. The pharynx is widest opposite the first branchial arch, behind which it narrows rapidly. Behind, or rather between the branchial arches of the fourth pair, the pharynx passes back into the œsophagus, a narrow tube from the ventral border of which the lungs are already commencing to arise as lateral diverticula.

The anterior part of the body, immediately behind the constricted neck, presents a rounded swelling on each side, caused by the head kidney.

3. THE HEART AND PERICARDIAL CAVITY.

The pericardial cavity lies in the floor of the pharynx, in front of the commencing liver, and opposite the four pairs of branchial arches. Its side walls are very thin, as also is its floor except where it is thickened in front by the sucker (Fig. 4, S). In sagittal section it is conical in shape, with the apex anterior. It is in communication behind with the general body cavity, which at this stage consists, owing to the great bulk of the food-yolk, of very narrow chink-like spaces on either side of the body.

The pericardial cavity is lined by a single layer of flattened epithelial cells. Its dorsal part is divided into right and left halves by a median mesocardial septum, the mode of development of which has been explained in the preceding section.

The heart (Plate XIV., Fig. 4) is a tube slightly twisted on itself, and slung to the dorsal wall of the pericardium by the mesocardial septum. Its walls consist of two layers of cells: an outer thick layer derived from the splanchnic mesoblast, and an inner layer of flattened cells formed by the cardiac endothelium, the origin of which has been discussed above. The heart is, in preserved specimens, sometimes filled with a coagulum, but no blood corpuscles are yet present in it, unless a few isolated cells, very similar in appearance to the endothelial cells, though rather more spherical in shape, and which in some specimens lie in the cavity of the heart, are to be regarded as corpuscles.

Posteriorly the heart is formed by the union of two large vitelline or hepatic veins (Fig. 4, V H), which run forwards from the lateral parts of the yolk mass and by the sides of the hepatic diverticulum. These unite in the median line to form what will afterwards be the sinus venosus, into which open the two Cuvierian veins, V D, which at present are of very small size. In front of the sinus venosus the heart runs forwards and slightly to the left side as the auricular portion, then bends over to the right side to form the ventricle, H V, and then back again towards the median plane to become the truncus arteriosus, which is in connection with the anterior wall of the pericardial cavity.

There is as yet no separation between the auricular and ventricular portions of the heart, but there is a sharp constriction between the ventricle and the truncus arteriosus. When the truncus arteriosus reaches the anterior wall of the pericardial cavity, it divides at once into right and left branches, which run directly outwards at right angles to the axis of the body, and just below the floor of the pharynx. Each division of the truncus is flattened dorso-ventrally and ends blindly, its extremity usually showing a more or less marked tendency to divide into two or three lobes.

4. THE BLOOD VESSELS OF THE VISCERAL ARCHES (Fig. 4).

a. The Mandibular Arch.—We have found no vessels at this stage that can be considered as belonging to the mandibular arch. The arch itself is hardly recognisable as such.

b. The Hyoid Arch.—Along almost the entire length of the hyoid arch, and occupying almost exactly its axis, there extends a narrow irregular lacunar space (Fig. 4, E H'). Its dorsal end reaches almost to the level of the aorta, A; its ventral end is difficult to trace, but can be followed with tolerable certainty as far as is shown in Fig. 4, and with more doubt right down to the floor of the pharynx. Along its whole length it has the appearance of an irregular lacunar space between the mesoblast cells, and possessing no true walls of its own. In places, however, the mesoblast cells immediately surrounding it are more closely arranged than elsewhere, as though tending to form a definite wall.

Opposite the dorsal end of this lacunar vessel, which we shall find represents the efferent branchial vessel of the hyoid arch, there is a very small diverticulum of the dorsal aorta (Fig. 4, E H). This lies to the inner side of the lacunar vessel, and has as yet no connection with it.

c. The First Branchial Arch.—In this arch two vessels are present, a posterior and smaller afferent vessel, and an anterior and larger efferent one.

The afferent vessel (Fig. 4, A F') is small in size and irregular in shape; it is widest at its dorsal end, which lies opposite the base of the external gill, and sends a small diverticulum outwards into the gill. This diverticulum is lacunar, *i.e.*, it has rather the appearance of an irregular channel between the mesoblast cells of the gill than of a vessel with definite walls. In some specimens the diverticulum ends blindly; in others it appears to be connected by very narrow lacunar channels with the efferent vessel. Below the base of the external gill the afferent vessel continues straight downwards in the side wall of the pharynx; it gradually diminishes in size, and ends blindly some little distance from the truncus arteriosus, with which, however, it is connected by a solid string of mesoblast cells, more definitely arranged, and staining more deeply than the general mesoblast of the arch.

The efferent vessel (Fig. 4, E F') lies anteriorly to the afferent, and is of much larger size. Its ventral end lies very close to the truncus arteriosus, and in a straight line with it; there is, however, a distinct interval between the two vessels, and the lower end of the efferent vessel is blind. Passing upwards in the wall of the pharynx the efferent vessel gets rapidly larger; it reaches its greatest size opposite the external gill, into which it extends, and within which in some specimens, as already noticed, it is connected with the afferent vessel by lacunar channels. Above the gill the efferent vessel narrows rather suddenly, and then continues upwards as a definite cylindrical vessel to the dorsal aorta, into which it opens (Fig. 4, A).

The efferent vessel has a well-developed wall of flattened epithelial cells; its cavity contains a few blood corpuscles, which also occur, though in smaller numbers, in the afferent vessel.

d. The Second Branchial Arch.—The vessels of this arch are very similar to those of the first branchial arch.

The afferent vessel (Fig. 4, A F″) is lacunar in character, and of very small size. It lies opposite the base of the second external gill, and communicates through one or more narrow intercellular passages with the efferent vessel; below the gill it extends a very short distance ventrally.

The efferent vessel (Fig. 4, E F″) is about the same size as that of the first branchial arch, and has very similar relations. It is widest opposite the external gill; below this point it narrows rather suddenly, and runs downwards in the arch to the level of the floor of the pharynx, where it ends blindly some little distance from the truncus. Above the gill it also narrows suddenly, and then continues upwards and inwards as a cylindrical vessel with more definite walls than before to the dorsal aorta, A, into which it opens. It contains blood corpuscles.

e. The Third Branchial Arch.—A small lacunar space or cleft (Fig. 4, E F‴), about the level of the external gills, is the only vessel present in this arch. It apparently represents the commencement of the efferent vessel of later stages.

f. The Fourth Branchial Arch.—In this there are as yet no vessels.

5. THE DORSAL AORTÆ.

Lying along the roof of the pharynx there are two aortæ, one on each side (Fig. 4, A). Opposite the hinder end of the pharynx they are almost parallel to each other, and lie just above the pharynx and a little ventral to the notochord. Passing forwards, the two aortæ diverge from each other very markedly; they are furthest apart opposite the first branchial arch, in front of which they converge again. Opposite the hyoid arch each divides into a dorsal and a ventral branch. The dorsal branches of the two sides (Fig. 4, A M′) run upwards and inwards, and meeting each other in the median plane, unite to form a commissural vessel immediately behind the infundibulum. The ventral branches, which are very slender, continue forwards, and end blindly close to the median plane, between the brain and the pharynx, and on a level with the eye-stalks.

Posteriorly to the division into dorsal and ventral branches, the aortæ are large vessels, somewhat compressed dorso-ventrally. The opening of the first branchial efferent vessel is wide, that of the second somewhat narrower.

Behind the pharynx the two aortæ approach each other, and about the level of the anterior nephrostomes they meet and unite to form the single dorsal aorta. This may be traced backwards, diminishing in size, as far as the posterior or third nephrostomes, beyond which level it does not extend at this stage.

Blood corpuscles occur in the dorsal aortæ, but are very few in number. About the level of the first nephrostomes the aortæ show slight bulgings (Fig. 4, G) on their ventral walls, which are the commencements of the glomeruli.

6. The Condition of the Blood Vessels at Stages Earlier than $4\frac{1}{2}$ mm.

It will be convenient if we give here the results of our observations on the condition of the blood vessels in tadpoles at stages earlier than $4\frac{1}{2}$ mm.

a. The Blood Vessels in 4 mm. Tadpoles.—At this stage there are no vessels in the visceral arches themselves. The dorsal aortæ are just commencing to form in the roof of the pharynx, each consisting at present of three independent lacunar spaces opposite the dorsal ends of the arches. Of these spaces, the first or most anterior one lies opposite the posterior part of the hyoid arch; the second or middle one is the largest, and extends from opposite the middle of the first branchial arch to the posterior part of the second branchial arch; the third space is small, and lies opposite the third branchial arch.

b. The Blood Vessels in $4\frac{1}{4}$ mm. Tadpoles.—The dorsal aorta is now a continuous vessel on each side, the three separate lacunæ of the earlier stage having opened out into one another. It now extends from the level of the infundibulum to the posterior end of the head. In the middle part of its course, opposite the hyoid and first two branchial arches, the aorta is now acquiring definite walls;

but at the two ends, and especially posteriorly, it is still a mere lacunar space in the mesoblast.

From the outer side of each aorta three diverticula are given off. The first of these passes into the hyoid arch, and is considerably larger than the aorta itself at this level; the second is rather smaller and passes into the first branchial arch; and the third diverticulum passes into the second branchial arch.

In addition to these diverticula of the aorta, small lacunar spaces are present in the dorsal part of the hyoid arch; a series of irregular detached lacunæ occur along the whole length of the first branchial arch, and a few small lacunar spaces near the dorsal end of the second branchial arch.

c. *The Mode of Formation of the Blood Vessels.*—It is difficult to determine the precise mode in which the blood vessels arise. In the first instance, as in the first appearance of the aorta, or of the efferent vessel of a branchial arch, a number of irregular and independent lacunar spaces are seen, which are apparently merely local enlargements of the intercellular spaces present everywhere in the mesoblast. These spaces rapidly enlarge, and open into one another; the mesoblast cells bounding them soon acquire distinctive characters; they become arranged more closely and more regularly; they stain more deeply than the other mesoblast cells, and they soon form a definite wall to the vessel.

The further growth of the vessel may either be effected by a continuance of this process, new lacunæ arising independently in the mesoblast, and then opening out into one another and into the already formed vessel; or it may take place in a different fashion, the mesoblast cells becoming arranged in compact strings continuing the lines of the already formed vessels, which strings are at first solid, but become tubular by the breaking down of their central or axial cells, and so become vessels. This second mode of growth appears to be the more usual; it is probably to be regarded as a modified form of the first method.

Blood corpuscles are at first absent; they, however, soon appear, though in small numbers. They form *in situ* in all parts of the body, and may be seen in lacunæ which still have no connection with other vessels. They are apparently in all cases individual mesoblast cells, and are at any rate in many cases formed by division of the cells which form the walls of the vessels. They are at first spherical, nucleated, and richly studded with yolk granules.

7. General Considerations.

The above description raises several questions of great interest, which it will be well to consider at once. The points of most importance appear to be the following:—

i. The heart is from the first in connection with the veins of the yolk mass; it is, indeed, formed by the union of these veins, and its endothelial lining is derived, in part at any rate, from the yolk cells.

ii. The heart has at first, and for some time, no connection with the branchial vessels. At $4\frac{1}{2}$ mm. the heart is considerably twisted on itself, and all its main divisions may already be recognised, the truncus arteriosus being, perhaps, the most clearly differentiated. Vessels are already present in the hyoid and the first three branchial arches, and in the first and second branchial arches have attained considerable development. None of those vessels, however, have any connection with the heart.

iii. The dorsal aortæ form very early in the roof of the pharynx, and soon acquire connection with the efferent branchial vessels: posteriorly the aortæ extend a very short distance behind the head.

iv. In the visceral arches the efferent blood vessels are the first to appear. At $4\frac{1}{2}$ mm. these are the only vessels present in the hyoid and the third branchial arches, where they have the form of irregular lacunar spaces, apparently intercellular in origin, and having as yet no connection with any other vessels. In the first and second branchial arches the efferent vessels are of much larger size, and extend almost the whole length of the arches; they are widest opposite the external gills, and are prolonged ventralwards almost to the truncus arteriosus, while dorsally they open into the aorta. The greater part of the length of the vessel is still lacunar, but the dorsal part, in communication with the aorta, is a well-formed tubular vessel, with walls formed of flattened epithelial cells. This distinction is important, because examination of earlier stages shows that this part of the vessel is formed as an outgrowth from the aorta towards the lacunar part of the efferent vessel, and not by extension of the lacunar vessel to meet the aorta.

The arrangement calls to mind the continuity between the vascular trunks and lymphatic spaces of Amphioxus, on which Lankester* has recently laid stress.

v. The afferent branchial vessels develope subsequently to the efferent vessels. At $4\frac{1}{2}$ mm. afferent vessels are present only in the first and second branchial arches, lying immediately behind the corresponding efferent vessels. The afferent vessel arises as an irregular lacunar space opposite the base of the external gill; it is at first independent, but soon acquires connection with the efferent vessel by narrow lacunar passages in the substance of the gill process, which passages give rise later on to the capillary loops of the gills. The ventral part of the afferent vessel extends downwards behind the efferent vessel, towards the truncus arteriosus, but does not at $4\frac{1}{2}$ mm. reach so far down as the efferent vessel.

vi. At $4\frac{1}{2}$ mm. there are very few blood corpuscles present. The heart in many specimens has none, but in others a few may be seen. The dorsal aorta may or may not contain corpuscles; but in the efferent branchial vessels a few are always present. These corpuscles are spherical nucleated cells, richly laden with yolk granules; they resemble very closely the cells of the surrounding tissues, and we have obtained direct evidence that in the efferent branchial vessels they are formed *in situ*, from the walls of the vessels themselves.

The description given by Maurer† of the development of the vessels in *Rana esculenta* differs in some important respects from the above account.

According to Maurer, the efferent branchial vessels appear as lacunar spaces in the connective tissue of their respective arches, and have at first no connection with the heart; while the dorsal aorta is not yet formed. The efferent vessels soon acquire communications with the truncus arteriosus, and a little later with the dorsal aorta. There is therefore a stage in which there is a single vessel in the arch, forming a direct communication between the heart and the aorta. As the development of the vessels commences earliest in the anterior arches, this stage is met with in succession in the arches from the first branchial to the fourth branchial. It will be

---

* Lankester, "Contributions to the knowledge of Amphioxus lanceolatus," "Quart. Journ. of Micros. Science," xxix., 1889, p. 379.

† Maurer, "Die Kiemen und ihre Gefässe bei Anuren und Urodelen Amphibien," "Morphologisches Jahrbuch," xiv., 1888, pp. 178-184.

convenient to postpone consideration of the vessels of the hyoid arch till a later stage.

The afferent branchial vessel of each arch, according to Maurer, arises from the ventral end of the corresponding efferent vessel close to the truncus arteriosus, and extends upwards along the posterior border of the efferent vessel until it reaches the base of the external gill; here it ends blindly for a time, but soon forms a loop which extends into the gill, and turning back opens into the efferent vessel at its widest part opposite the root of the gill.

The most important difference between this account and our own is that while Maurer finds a stage in which there is no afferent vessel in the arch, and the efferent vessel forms a direct communication between the heart and the aorta, we find no such stage at all, the afferent vessel being well developed before the branchial vessels have acquired any connection with the heart; while the connection, when it is formed at a later stage, is effected, as we shall describe further on, by the afferent, and not by the efferent vessel. A second point of difference is that, while Maurer describes the afferent vessel of each arch as arising from the ventral end of the efferent arch, and growing up dorsalwards to the gill, we find that its dorsal end is the first part to appear, opposite the gill, that it early acquires connection with the efferent vessel in the gill, and then grows down ventralwards towards the heart, meeting and fusing at a still later stage with an outgrowth from the truncus arteriosus.

These differences are perplexing and very unwelcome, and we have accordingly studied these stages repeatedly and with great care, and on a considerable number of specimens, in order to make sure of the correctness of our observations. The investigation is a troublesome one, for it is a most difficult matter to determine with satisfactory precision the mode of origin and the exact boundaries of channels which, like the branchial vessels, appear first as irregular lacunar spaces with no proper walls of their own. It is, therefore, quite possible that our account may require modification in minor details, but of its essential correctness we are convinced.

On the other hand, Maurer's observations have clearly been conducted with great care, and we see no ground for questioning the accuracy of his descriptions. We must therefore conclude that the early stages of formation of the branchial vessels are different in the species investigated by Maurer, *Rana esculenta*, and in *Rana temporaria*, which we have employed for our own observations.

In support of this view, which, at first sight, may not commend itself to morphologists, we would note the following points:—

i. In the tadpoles of *Rana temporaria* which we have examined, we have found a certain amount of variability in respect to the size and relations of these branchial vessels, though an agreement in all essential respects.

ii. In one specimen, but in one only out of a considerable number, we have found a condition of things resembling that described by Maurer. This tadpole measured $4\frac{1}{2}$ mm. in length, but is thicker, and slightly different in general shape to the ordinary tadpoles of the same length. We have no special record with regard to its origin, but can hardly doubt that, while slightly abnormal, it is really a tadpole of *Rana temporaria*.

In Fig. 5 a transverse section of this tadpole is given, passing on the left side through the first branchial arch, and on the right side through the first branchial pouch. The figure is diagrammatic, being compounded from three or four successive sections, so as to show the entire length of the branchial vessels.

The efferent vessel is seen here to form a continuous trunk connecting together the truncus arteriosus and the aorta exactly as described by Maurer. The middle portion of the vessel, opposite the external gill, is wide and lacunar, while the dorsal and ventral portions are cylindrical, with true epithelial walls. The afferent vessel is present, but does not agree with Maurer's description, for while it lies very close to the efferent vessel in the gill process, its ventral end has no communication with the truncus, but lies some distance from it.

This is the only tadpole of this age in which we have found the efferent vessel forming a direct connection between the heart and the aorta.

iii. We have already noticed that there are strong independent reasons for thinking that the early development of the frog has been considerably modified, and cannot be taken as affording a correct view of its ancestral history. The branchial vessels afford an additional illustration, for it is clear that neither Maurer's account of their development, nor our own, could possibly be interpreted as strict recapitulation. It is inconceivable that an ancestor of frogs should have been an animal with a brain, a notochord, and a heart with the characteristic Vertebrate twist, and with the main subdivisions already indicated, and yet that this heart should have had no connection whatever with the branchial vessels or the aorta.

We are as yet completely in the dark as to the steps by which the branchial respiration, and consequently the branchial circulation, of Vertebrates were first acquired. The development of the branchial vessels in fishes has been as yet very imperfectly studied, and we shall probably be wise in exercising great caution in attempting to reconstruct the past ancestral history from the development of these vessels in higher Vertebrates. It is commonly assumed that the condition in which there is in each arch a continuous vessel connecting the heart and aorta directly is an ancestral one; and if this be the case—which is by no means proved—then the mode of development described by Maurer in *Rana esculenta* would be more primitive than that observed by us in *Rana temporaria*; and the exceptional tadpole of *R. temporaria* which we have described above, could then be explained as a case of reversion to the more ancestral type.

We have thought it best to introduce this somewhat lengthy discussion in this place, because it arises directly from our observations on the earlier stages, and because the later stages are of little interest except in connection with the questions we have just considered. We will now return to the description of these later stages, commencing with the tadpole at the stage when it emerges from the spawn and becomes free.

III. THE CONDITION OF THE HEART AND BLOOD VESSELS IN 5 MM. TADPOLES.

1. EXTERNAL CHARACTERS.

As compared with the earlier stage, the newly hatched tadpole differs chiefly in the greater size of the tail, to which the increased length of the animal is entirely due, and which now forms about one-third of the entire length. The abdomen is rather less bulky, owing to the absorption of yolk for food. The sucker is very prominent, and larger than before. The olfactory pit has a rather narrower mouth; and the longitudinal ridge, prolonged backwards from its hinder border and marking the position of the lens of the eye, is rather more prominent. The position of the ear can be recognised as a slight swelling on the surface. The two external gills of each side are larger; that on the first branchial arch is notched at its free posterior border into three blunt lobes, and that of the second arch into either two or three lobes. The third branchial arch has no external gill as yet. The hyoid arch is more clearly marked out than before, and slight vertical grooves indicate the positions of the future gill clefts.

2. INTERNAL ANATOMY.

The internal structure differs but little from that of the $4\frac{1}{2}$ mm. tadpole. There is, as might be expected, a certain amount of variability; tadpoles of the same actual length not agreeing precisely in their anatomy. In a certain number of specimens we have noticed distinct asymmetry, the right side of the animal being slightly in advance of the left in its development; but whether this difference is constant we have failed to determine.

The fourth branchial arch is now more clearly defined than before; and the several gill pouches have begun to open out, the two layers of hypoblast that form their walls separating slightly from each other, so that the cavity of the pharynx now extends outwards as a narrow cleft a short distance along each diverticulum.

The lungs are slightly larger than before, but are still very small.

3. THE HEART AND PERICARDIAL CAVITY.

The pericardial cavity is much as before. Its floor and sides are lined by a thin layer of flattened epithelial cells; its posterior wall is formed by the anterior surface of the liver diverticulum. Below and at the sides of the liver the mesoblast is split into somatic and splanchnic layers, and the cœlomic space between them, which is still very narrow, opens into the pericardial cavity.

The twisting of the heart (Plate XIV., Fig. 6) has increased very considerably. From the sinus venosus the heart runs forwards to the left side of the pericardial cavity, then bends sharply downwards, and runs transversely across to the right side, and then bends sharply forwards, upwards, and to the left, reaching the anterior wall of the pericardial cavity in the median plane. The first of the two bends forms the auricular portion of the heart, H A, and the second the ventricular, H V. The ventricle lies at a level slightly ventral to the auricle, and is separated from the truncus arteriosus, in front, by a sharply marked constriction.

Outside the pericardial cavity the truncus arteriosus divides at once into right and left branches, which are much wider than at the earlier stage, and run directly outwards, beneath the pharynx. Each branch gives off from its anterior border a small blind diverticulum into the hyoid arch (Fig. 6, A H), and then divides into afferent trunks for the first and second branchial arches (Fig. 6, A F.1, A F.2), which run outwards a short distance in these arches, and end blindly.

The heart now contains blood corpuscles, which vary greatly in number in different specimens. In some they are absent, or nearly so, while in others they are abundant. They are all spherical in shape, and loaded with yolk granules. In the hepatic veins the blood corpuscles may be seen separating from the walls of the veins.

4. The Blood Vessels of the Visceral Arches.

a. The Mandibular Arch.—An extremely small diverticulum from the front part of the dorsal aorta is present at this stage just behind the level of the infundibulum (Fig. 6, E M). Though extremely small, not exceeding in length the diameter of the aorta at its point of origin, we have found this vessel to be constantly present. Its relations are exactly those of the dorsal portions of the hinder efferent vessels, and we believe it should be regarded as the dorsal or aortic portion of an efferent branchial vessel belonging to the mandibular arch. Its later development and its homologies will be considered afterwards.

In the ventral part of the mandibular arch a lacunar vessel (Fig. 6, V M) is present, which communicates below with an irregular longitudinal sinus lying in the floor of the pharynx, dorsal to the sucker. This mandibular vein, V M, exactly corresponds to the ventral part of the efferent vessel of the hyoid arch, V Y, and may reasonably be regarded as part of an efferent vessel belonging to the mandibular arch.

b. The Hyoid Arch.—The condition of the vessels in the hyoid arch is very instructive. From the lateral branch of the truncus arteriosus, just before its division into the afferent trunks for the first and second branchial arches, a short, anteriorly directed diverticulum arises (Fig. 6, A H). Its length is about equal to the diameter of the trunk for the first branchial arch, and it ends blindly. Its position and relations inevitably suggest that it is a true afferent trunk; and inasmuch as it lies just anterior to the trunk for the first branchial arch, it seems justifiable to refer it to the hyoid arch, especially as the other elements of a branchial vessel are present in this arch.

These other elements consist, firstly, of the lacunar vessel, which at $4\frac{1}{2}$ mm. extended along the whole length of the arch (Fig. 4, E H'). At the present stage, 5 mm., the middle portion of this vessel has become obliterated, so that the vessel is divided into two separate portions, of which one (Fig. 6, E H') lies in the dorsal part of the arch, not far from the dorsal aorta, while the other (Fig. 6, V Y) lies in the ventral part of the arch. This ventral portion, V Y, which may be conveniently spoken of as the hyoid vein, opens below, as shown in Fig. 6, into the longitudinal sinus above the sucker, into which, as described above, the mandibular vein, V.M, opens further forwards.

Besides this lacunar vessel, which clearly corresponds to the efferent vessel of a branchial arch, there is also present, in the dorsal part of the arch, a diverticulum from the dorsal aorta (Fig. 6, E H). Though still a short vessel, this diverticulum is considerably larger than at $4\frac{1}{2}$ mm. (cf. Fig. 6, E H). Its distal or ventral end is blind, and lies very close to the dorsal end of the hyoid efferent vessel, E H'. In some, but not all specimens, at a slightly later stage, we have found the two vessels meeting and opening into each other.

The hyoid arch, therefore, possesses at 5 mm. all the vascular elements found in the branchial arches excepting the afferent branchial vessel. These elements show unmistakeable signs of degeneration. The vascular arch is never completed; the efferent lacunar vessel has already become obliterated in the middle part of its course; the diverticulum from the truncus arteriosus is very small, it never becomes connected with the other vessels of the arch, and it disappears altogether at a slightly later stage; and the diverticulum from the aorta, though rather better developed than that from the truncus arteriosus, also persists for but a short time, and at the next stage ($6\frac{1}{2}$ mm.) has almost disappeared. These degenerative changes may reasonably be associated with the absence of a gill on the hyoid arch, which will also explain the failure to develope an afferent branchial vessel for this arch.

From the accounts given by Maurer and others, there seem to be very great differences in the condition of the vessels of the hyoid and mandibular arches met with amongst Amphibia themselves, and it is hardly possible to determine which of these conditions is to be regarded as the most primitive for the group. It is instructive, however, to note that in the frog the vessels of these arches are in a more degenerate condition than in the Amniote Vertebrates, in which these vessels form continuous arches between the heart and aorta in the early stages of development.

c. The First Branchial Arch.—The afferent branch from the truncus arteriosus (Fig. 6, A F.1) has already been described. It is a vessel of fair size and with well-formed walls, which runs transversely outwards beneath the pharynx, and ends blindly about the junction of its floor and sides. The efferent branchial vessel, E F′, is very large opposite the external gill, where it lies in the middle of the arch. From this point it extends downwards in the anterior part of the arch towards the ventral surface, narrowing as it does so, and ending blindly at the lower part of the side of the pharynx, in a line with, but a little distance from, the blind end of the afferent branch from the truncus arteriosus. Above the gill the efferent vessel, as in the earlier stage, narrows considerably, becomes a tubular vessel with well-formed epithelial walls, and opens into the dorsal aorta opposite the widest part of the pharynx.

The afferent branchial vessel (Fig. 6, A F′) has much the same relations as in the earlier stage; it is still lacunar, and extends downwards towards the truncus arteriosus somewhat further than before, but does not yet meet it. The dorsal end of the afferent vessel communicates with the efferent vessel by one well-developed capillary loop in the external gill; and ventral to this loop the afferent and efferent vessels are in direct communication by very narrow and irregular lacunar passages.

The efferent branchial vessel contains numerous blood corpuscles, apparently budded off from its walls.

d. The Second Branchial Arch is very similar to the first as regards its vessels, though these are of slightly smaller size, and less fully developed.

The afferent branch from the truncus arteriosus (Fig. 6, A F.2) is posterior and slightly ventral to that for the first arch. The efferent branchial vessel (E F″) does not extend so far down the arch as does that of the first arch; opposite the gill it gives off a diverticulum which enters the gill, but ends blindly close to the afferent branchial vessel, but without reaching this. The afferent branchial vessel (A F″) is still a small irregular lacunar space, which in some specimens communicates with the efferent vessel opposite the gill, but in others ends blindly; its ventral portion extends downwards towards the truncus arteriosus, but ends blindly some distance from it.

e. The Third Branchial Arch has vessels which agree almost exactly with those of the hyoid arch. From the posterior border of the branch of the truncus arteriosus to the second branchial arch, just after this latter has separated from the branch for the first arch, a small backwardly-directed diverticulum arises (Fig. 6, A F.3), which is shown by its later development to be the commencement of the afferent branch for the third arch; it is at this stage extremely short, and ends blindly.

The efferent branchial vessel (E F‴) is represented as before by a small irregular lacunar space in the third branchial arch at the level of the external gills, which at present has no communication with any other vessel.

From the dorsal aorta a small and short blind diverticulum (E F.3) extends outwards and downwards in the arch. It is directed straight towards the lacunar efferent vessel, but stops short some distance from it.

f. The Fourth Branchial Arch has as yet no vessels.

g. The Dorsal Aorta.—The dorsal aortæ in the roof of the pharynx have much the same relations as before; they are furthest apart opposite the gills, in front of which level they bend in towards each other rather sharply, at the same time narrowing considerably. They nearly meet just behind the infundibulum, where they are connected by the posterior infundibular commissural vessel (Fig. 6, A M′), and then continue forwards along the sides of the brain, between it and the eyes. Behind the gills the aortæ bend towards the median plane, at first sharply, and then more gradually; the two vessels meet a little behind the level of the first nephrostomes, and unite to form the median dorsal aorta, which extends back beneath the notochord somewhat further than at the earlier stage.

Above the pharynx each dorsal aorta receives five vessels; of these the first, second, and fifth (E M, E H, and E F.3), belonging respectively to the mandibular, hyoidean, and third branchial arches, are small, short, and blind distally; while the other two (E F.1 and E F.2) are of large size, and, as already noticed, return to the aorta the blood from the gills on the first and second branchial arches.

Shortly before the two aortæ unite to form the median dorsal aorta, each is in connection with a glomerulus, G, which is now a well-marked and sacculated diverticulum from the ventral and outer surface of the aorta.

h. The Veins.—The liver is now assuming more definite form, and the hepatic and vitelline veins are distinct from one another. Of the hepatic veins (Fig. 6, V H) the right is apparently constantly of larger size than the left. The Cuvierian veins, V D, are considerably larger than before, and each divides dorsally into anterior and posterior cardinal veins, of which the latter are of large size, and surround the tubules of the head kidney in an intricate fashion.

5. GENERAL CONSIDERATIONS.

There is little to call special attention to at this stage, except the very significant arrangement of the vessels in the mandibular, hyoidean, and first branchial arches. In the mandibular arch the vessels are just forming, and are now in a condition comparable to that of the hyoid vessels at $4\frac{1}{2}$ mm. In the hyoid arch the vessels have reached their full development, and have already begun to undergo degenerative changes. In the third branchial arch the actual condition is closely comparable to that of the hyoid arch. A lacunar efferent vessel is present, as also are diverticula from the aorta and truncus arteriosus, these three elements being perfectly independent, and indeed widely separate. While, however, the hyoidean vessels are commencing to degenerate, those of the third branchial arch advance steadily in development during the succeeding stages.

The condition of the heart is worthy of notice, especially if it be remembered that as yet the truncus arteriosus ends blindly in front.

IV. THE CONDITION OF THE HEART AND BLOOD VESSELS IN 6½ MM. TADPOLES.

1. EXTERNAL CHARACTERS.

As in the previous stage, the increase in length is due almost entirely to growth of the tail, which now forms nearly half the entire length of the tadpole. Apart from this, the chief changes are as follows:—The body is much more slender, and the ventral surface less protuberant, owing to continued absorption of the yolk. The dorsal fin is much more prominent, and extends forwards to the hinder end of the head; the ventral fin also is better developed than before, beginning immediately behind the yolk mass, and having the cloacal aperture on its ventral edge.

In the head, the sucker is of large size, as before; the stomatodæal depression is deeper and better defined than at the earlier stages, and is overhung by a distinct upper lip. The external gills of the first and second branchial arches are much larger than before; they are directed outwards and backwards, the first partially overlapping the second, which is placed rather more dorsally. Each gill is deeply notched along its posterior border into four to five lobes, which decrease in size from above downwards; the uppermost lobe is much the largest, and gives off minor lobes along its posterior or inner surface. External gills are present as small unbranched processes on the third branchial arch as well, but are overlapped and concealed by the gills of the first and second arches when the animal is viewed from the side.

The head is separated from the body by a distinct constriction, immediately behind which a rounded swelling is formed by the head kidney on each side.

Though the mouth is not yet open, the tadpole shows a distinct increase in bulk as compared with the 5 mm. stage. It has occurred to us as possible that the suckers may be used for absorbing food, and that in this way the increase may be explained. Tadpoles up to about this age adhere by their suckers to the gelatinous mass of the spawn from which they have recently emerged; and sections of the sucker show that the greatly elongated and columnar cells of the sensory layer of the epiblast covering them are often produced at their free ends into protoplasmic processes, that would seem well fitted for absorbing the jelly. We have also found the pharynx at this and at earlier stages to be very commonly completely filled, in hardened specimens, with a dense and coherent coagulum, the material for which may be obtained in the way suggested above.

2. INTERNAL ANATOMY.

The pharynx is still of large size, being wide both laterally and dorso-ventrally (Figs. 8, 9, 10, P). It is, however, relatively to the whole head, smaller than at 5 mm., owing to the increased size of the brain. In front of the gill clefts the pharynx is compressed laterally, and prolonged forwards beneath the brain to the septum dividing it from the stomatodæal depression. At this septum (Fig. 10, S T) the epiblast and hypoblast are in close contact with each other, the hypoblast and the deeper or nervous layer of the epiblast fusing indistinguishably with each other.

At its hinder end, opposite the ears, the pharynx narrows rapidly and passes into the œsophagus, which has a very small lumen, appearing in transverse sections as a vertical slit.

The gill slits (Fig. 9) are much as in the previous stage, the two layers of hypoblast bounding each slit being still in contact along almost their whole width. Along the outer edges of the gill slits the hypoblast cells fuse completely with the deeper or nervous layer of the surface epiblast. Well marked but shallow grooves of the epiblast mark the position of the gill clefts on the surface. The hyomandibular cleft (Fig. 9, C H) is very similar to the hinder or branchial clefts, but is connected with the surface epiblast for a shorter distance than these.

The thyroid body is present as a well-marked median diverticulum of the ventral wall of the pharynx, near its anterior end. It is directed backwards and very slightly downwards, and does not quite reach the front wall of the pericardial cavity.

The lungs are more prominent than before, as lateral outgrowths from the ventral wall of the œsophagus in the neck region. The inner surface of their lining epithelium is strongly pigmented.

3. The Heart and Pericardial Cavity.

The pericardial cavity has much the same relations as before, but is of smaller size, owing partly to the increased dimensions of the heart, and partly to extension forwards of the liver. The cavity communicates with the cœlom behind, and also dorsally; the dorsal part of the cœlom on either side, into which the nephrostomes open, being prolonged forwards as a narrow vertical channel, which communicates just in front of the Cuvierian veins with the dorsal and posterior end of the pericardial cavity.

The heart has the same general relations as before, but the vertical twist is more marked (Fig. 7), the auricular portion lying distinctly dorsal to the ventricular. The distinction between auricle and ventricle is still very slightly marked. The walls of the truncus arteriosus are rather thicker than before.

The truncus arteriosus leaves the pericardial cavity in the median plane in front, and immediately divides into right and left branches, which run horizontally outwards (Fig. 10). Each branch divides about the level of the side wall of the pericardial cavity, and about midway between the median plane and the surface, into two vessels, the anterior of which (Fig. 10, A F.1) runs directly outwards to the first branchial arch, while the posterior runs backwards and outwards, and divides to supply the second and third branchial arches (Fig. 10, A F.2, A F.3).

4. The Blood Vessels of the Visceral Arches (Figs. 7–10).

a. The Mandibular Arch.—The diverticulum of the dorsal aorta that was described as a very minute structure in the 5 mm. tadpole, is present, and has increased considerably in length (Fig. 7, E M). It runs downwards, outwards, and slightly forwards, just in front of the hyomandibular cleft; and after a short course dilates suddenly, or rather opens into a large vessel with sacculated walls, the mandibular vein (Fig. 7, V M), which runs downwards to the floor of the pharynx, where it opens into an irregular mesh-work of large lacunar channels, lying just above the sucker, and at the sides of the thyroid body.

The vessels of the mandibular arch are peculiar in several respects. Still we believe that they may be compared with tolerable certainty with those of the hyoidean and branchial arches. The diverticulum from the dorsal aorta (Figs. 6, 7, and 9, **E M**) must certainly be regarded as homologous with the similar diverticula in the hinder arches, as in relations and mode of development it is identical with these. The lacunar vessel (Figs. 6 and 7, **V M**), which we have termed mandibular vein, differs markedly from the efferent lacunar vessels of the branchial arches in appearing first in the ventral part of the arch (Fig. 6), instead of at the level of the external gills; it differs further in being from the first in connection at its ventral end with the lacunar sinuses above the sucker. The condition of the hyoid arch appears, however, to afford the clue. At 5 mm. the vessels, **V Y** and **V M** (Fig. 6), lying in the ventral parts of the hyoid and mandibular arches respectively, have precisely similar relations. The vessel, **V Y**, is shown by the $4\frac{1}{2}$ mm. stage (Fig. 4) to be the ventral end of the hyoid efferent vessel. It seems reasonable, therefore, to regard the mandibular vein, **V M**, as the efferent vessel of the mandibular arch. The fact that at $6\frac{1}{2}$ mm. (Fig. 7) the mandibular vein has extended dorsalwards, and is directly and unmistakeably continuous with the aortic diverticulum, **E M**, affords evidence of the strongest kind in support of this determination.

On this view, the chief peculiarity of the vessels of the mandibular arch is that, instead of the dorsal part of the efferent vessel being the first to form, it is the ventral end that developes earliest; and this peculiarity may fairly be ascribed to the absence of a gill, and to the presence and physiological importance of the sucker, with which the mandibular vein is in close relation.

It may be further noted that there is no room for doubt as to these vessels belonging to the mandibular arch, inasmuch as they lie clearly and distinctly in front of the hyomandibular cleft (*cf.* Fig. 9, **E M**).

From the relations of the vessels, it follows that the flow of blood must be from the dorsal aorta ventralwards towards the sucker.

b. The Hyoid Arch.—The small diverticulum of the truncus arteriosus that was present at 5 mm. has disappeared entirely, no vestige of it remaining. The diverticulum from the dorsal aorta is still present, but is much shorter than before, and ends blindly, (Fig. 7, **E H**).

The only other vessel that belongs to the arch is a large sacculated vein in the ventral half of the arch (Fig. 7, **V Y**), which commences with a narrow blind end about the level of the middle of the pharynx, and runs downwards, enlarging rapidly, and opening at its ventral end into the lacunar spaces above the sucker, with which we have just seen the mandibular vessel to be also connected. This vessel, V Y, may be spoken of as the hyoidean vein, and clearly corresponds to the vessel, V Y, of the 5 mm. stage (Fig. 6). The degenerative changes in the vessels of the hyoid arch, that had already commenced at 5 mm., have thus progressed considerably. The dorsal part of the efferent lacunar vessel (Fig. 6, E H') has disappeared completely, and the aortic diverticulum, E H, has undergone reduction in size.

c. The First Branchial Arch.—The afferent vessel of this arch is, as noticed above, the anterior of the two vessels into which the truncus arteriosus divides on each side (Figs. 7 and 10, A F.1). It runs at first directly outwards, then outwards and upwards in the first branchial arch; it is at first small, but opposite the base of the gill dilates rather suddenly. From this dilated part branches are given off into each lobe of the gill, the branch of each lobe running along its posterior surface to its tip, where it opens into the efferent vessel (Fig. 10, A F.1, E F.1). The afferent and efferent vessels of each lobe are also connected directly by small capillary loops which lie in the minor processes arising from the lobe (Figs. 8, 9, 10). The efferent vessel, E F', lies immediately in front of the afferent; it is wide opposite the gill, and widest of all opposite its uppermost or dorsal lobe. Above this level it narrows very considerably, and continues its course upwards and inwards until it reaches and opens into the dorsal aorta. Ventrally to the gill the efferent vessel narrows considerably, and is continued inwards in the floor of the pharynx towards the median plane. Along this part of its course it lies in close contact with the anterior wall of the afferent vessel. The efferent vessel extends inwards almost to the level of the division of the truncus into the afferent branches for the first and second branchial arches (Figs. 8 and 10, E F.1); it ends blindly, and has at this stage no connection with the afferent vessel at this point. The relations of the afferent and efferent vessels of this arch are shown somewhat diagrammatically in Fig. 8, where the greater part of the afferent vessel, A F.1, has been removed on the left side; the figure shows the base only of the external gill, and the capillary loops, G C, connecting the afferent and efferent vessels are diagrammatically expressed. The gill in its true proportions is seen in Figs. 9 and 10.

d. The Second Branchial Arch.—In this arch the vessels have essentially the same arrangement as in the first branchial arch, and there is no occasion to describe them in detail. The efferent vessel is prolonged ventralwards along the anterior border of the afferent vessel, but, as in the first arch, it ends blindly, and there is no communication between the afferent and efferent vessels of the arch except through the gill capillaries.

e. The Third Branchial Arch.—From the posterior border of the afferent vessel of the second arch, some distance beyond the point where it leaves the vessel of the first arch, but before it reaches the gill, a short afferent diverticulum (Figs. 7 and 10, A F.3) is given off, which enters the third branchial arch, and then ends blindly.

In the arch itself, at the side of the pharynx, there is an efferent vessel of considerable size (Fig. 7, E F'''), which is widest opposite the small external gill, and is prolonged upwards above this point to open into the dorsal aorta.

Opening into the efferent vessel opposite the gill is a small afferent vessel which is continued downwards in the arch, lying immediately behind the efferent vessel, but not extending so far ventralwards.

The condition of the external gill of the third branchial arch is somewhat variable in specimens of this size. When fairly well developed, as in Fig. 9, it contains a capillary loop connecting the afferent and efferent vessels together.

f. The Fourth Branchial Arch.—There is as yet no afferent diverticulum of the truncus arteriosus for this arch. An efferent vessel of some size is present in the upper part of the arch (Fig. 7, E F''''), which is prolonged upwards and forwards to join the dorsal aorta, just behind the point of opening of the efferent vessel of the third arch. From the widest part of the fourth efferent vessel a small but distinct pulmonary artery arises (Figs. 7 and 9, A P), which runs backwards and inwards to the rudimentary lung.

Behind the efferent vessel, and ventral to the origin of the pulmonary artery, a small lacunar space is present (Fig. 7, A F''''), which is apparently the first trace of the afferent vessel of the fourth arch, and which at present has no communication with any other vessel.

g. The Dorsal Aorta.—The course and relations of the dorsal aortæ are shown on the right side in Fig. 7, and from above in Fig. 9. In the latter figure the gills and gill arches of the right side are drawn from a section at a level somewhat ventral to that represented on the left side, where the entire course of the aorta is shown. The two aortæ are furthest apart opposite the first branchial cleft; in front of this level they converge very rapidly to the hinder border of the infundibulum; here they turn sharply forwards and slightly outwards, and run as the internal carotid arteries along the sides of the brain, passing dorsal to the eye stalks. The two aortæ are connected together by two transverse commissural vessels, one lying above and one below the infundibulum (Figs. 7 and 9, A M, A M'). Behind the first branchial clefts the two aortæ run backwards along the roof of the pharynx (Fig. 10, A). They converge at first rapidly, then more gradually, and unite about opposite the level of the first pair of nephrostomes of the head kidneys (Fig. 9). Immediately behind the point of union there is present on each side of the aorta a large glomerulus (Fig. 9, G), lying opposite almost the whole length of the head kidney.

Each dorsal aorta receives in the pharyngeal region, as already noticed, six efferent branchial vessels. Of these the first or mandibular vessel (Figs. 7 and 9, E M) is very slender, is in communication distally with the mandibular vein (Fig. 7, V M), and has no corresponding afferent vessel. The second or hyoid efferent vessel (Figs. 7 and 9, E H) is at this stage an extremely short blind diverticulum. The third and fourth are the large efferent vessels of the first and second branchial arches, which transmit to the aorta the blood from the large external gills of these arches, and which are in connection with the heart through the capillaries of these gills.

The last two—the fifth and sixth—are the efferent vessels of the third and fourth branchial arches, which at present have no communication with the heart. The pulmonary artery is present at this stage, but if any blood flows along it, this must be blood from the dorsal aorta.

h. The Veins.—We have but little to say at this stage concerning the veins, which we have studied in far less detail than the arteries.

The Cuvierian veins are larger than before, and now considerably exceed the hepatic veins in size. Of their two components, the posterior cardinal veins or sinuses are greatly dilated in relation with the head kidneys; while the anterior cardinals or jugular veins are also of large size, extending forwards along the sides of the head about the level of the notochord (Fig. 10, V R).

The great system of lacunar venous sinuses on the ventral surface of the head, in front of the pericardium and above the sucker, has already been alluded to; and the mandibular and hyoidean veins opening into these have been described (Fig. 7). From these large sinuses the blood is returned to the heart by the inferior jugular veins, a pair of large vessels which run backwards along the side walls of the pericardium, and open into the Cuvierian veins just before these join the sinus venosus.

In many respects the arrangement and proportions of the veins of the tadpole at this stage bear a striking resemblance to those of an adult Elasmobranch, such as a dogfish. In this connection we may note more especially the position and relations of the mandibular vein, and its communication with the heart through the inferior jugular vein, and also the enormous size of the front part of the posterior cardinal sinuses. The arrangement and proportions of the arteries of the head also support this comparison strongly.

5. General Considerations.

The most important points of a general nature that call for consideration at this stage are connected with the vessels of the mandibular and hyoidean arches. These can be only partially dealt with at present, and we propose to defer their discussion until we have described tadpoles of the next two stages—9 mm. and 12 mm. respectively.

The exceedingly early period at which the pulmonary artery developes is rather surprising, and deserves attention.

V. THE CONDITION OF THE HEART AND BLOOD VESSELS IN 9 mm. TADPOLES.

1. External Characters.

This is an interesting and well-marked stage, as it is at this period that the mouth-opening is first formed, by perforation of the septum between the stomatodæum and pharynx, and the tadpole first commences to eat. As might be expected, the exact time at which the perforation is effected is subject to some variation; but in all tadpoles that we have examined measuring 8½ mm. or under, the septum is still intact, while in many 9 mm. specimens, and all 9½ mm. ones, perforation has occurred, as shown in Plate XV., Figs. 11 and 12, z.

As regards external characters, the tail has increased considerably in length, and the body has become much more slender owing to absorption of the yolk, the diminution in bulk being very apparent when the animal is viewed from the ventral surface. The dorsal fin is large, and extends forwards to about the level of the ear. There is a well-marked rectal papilla, opening at the edge of the ventral fin, the cloacal opening being almost exactly at the middle of the length of the tadpole. The head kidney swellings are still very prominent, and immediately in front of them is the laterally constricted neck.

The head is broad, rounded anteriorly, and widest opposite the gills. The olfactory sacs are conspicuous, their depressions being deeper than at the former stages, and their posterior lips being particularly well marked. Both eye and ear are visible on the surface as small swellings, circular in outline. The mouth is a deep depression on the under surface of the head, a little distance from its anterior end; it is transversely elongated, and bounded by distinct lips, within which lie the horny jaws. The sucker on the under surface of the head is still single, and of the same shape as before, but is distinctly smaller.

The external gills of the first and second branchial arches are greatly developed, and about equal in length to the transverse diameter of the head at their base (Fig. 12). In the living animal the gills are carried projecting outwards and backwards from the head at an angle of about 45° with the axis of the body. Each gill has distinct muscles of its own, by means of which the entire gill or its several lobes can be moved freely and independently. Each gill consists of from five to seven main lobes, decreasing in size from above downwards, and each lobe gives off minor lobes along its posterior border. The circulation of the blood in the gills can be well studied in the living animal. Each lobe, and each of its minor lobes, contains two vessels, afferent and efferent, which lie close alongside each other, and communicate directly at the tip of the lobe (Fig. 12), the afferent vessel being placed posteriorly and somewhat ventrally to the efferent vessel. The third branchial arch bears an external gill which is much smaller than those of the first and second arches, and almost concealed by these. The heart's beats are rapid—about fifty to the minute. The whole surface of the tadpole is, as in the earlier stages, ciliated; the cilia working from head to tail, causing the animal when perfectly quiet to move forwards slowly in the water. Distinct opercular folds (Fig. 12, O F) are present on the ventral surface of the head, growing backwards from the posterior borders of the hyoid arches.

2. INTERNAL ANATOMY.

The proportions and relations of the organs in the anterior part of the head have changed very considerably as compared with the earlier stages, as will be seen on comparing Figs. 7 and 11. The brain has grown forward very markedly, the cerebral hemispheres being now conspicuous structures. The anterior lip has grown still more rapidly, so that the brain, in spite of its forward extension, no longer reaches to the anterior end of the body. This growth of the lips causes considerable deepening of the stomatodæal depression.

From the bottom of each olfactory pit a solid extension of the nasal epithelium has grown downwards towards the mouth, which it nearly meets just opposite to the septum. A slight diverticulum of the stomatodæum immediately in front of the septum extends towards it, but does not yet meet it; these two structures together forming the rudiment of the posterior narial passage.

The pharynx is very wide from side to side, widest opposite the second branchial arches; but is shallower dorso-ventrally than before. The second and third branchial clefts (Fig. 12, C B.2, C B.3) now open to the exterior, the opening being effected by separation of the two hypoblastic lamellæ of which each gill pouch consists, completed by a very slight groove-like depression of the surface epiblast. The openings are narrow slits lying immediately behind the first and second external gills respectively, and rather on the ventral surface. The opening of these two clefts to the exterior we have found to occur as nearly as possible simultaneously with the formation of the mouth by perforation of the stomatodæal septum.

The first branchial cleft, between the hyoidean and first branchial arches, extends at this stage very close to the surface, but does not actually open to the exterior.

The thyroid body arises from the bottom of a slight depression in the floor of the mouth as a slender solid rod of cells, which runs backwards and downwards beneath the floor of the pharynx, and enlarges posteriorly into a club-shaped solid body, reaching almost to the anterior wall of the pericardial cavity, but not quite touching it.

Immediately behind the last or fourth branchial cleft, the pharynx narrows very suddenly to form the œsophagus, which runs back through the constricted neck. The walls of the œsophagus are thick; the lumen is very small, and for a short distance is completely absent, reappearing further back as a vertical slit, from the ventral portion of which the lungs arise as a pair of lateral diverticula. This solid condition of the œsophagus we have found to be a constant feature in tadpoles of from $7\frac{1}{2}$ mm. to about $10\frac{1}{2}$ mm. in length. The lumen is not re-established until some time after the mouth perforation is completed; and it is for a time exceedingly narrow.

A solid œsophagus was first described by Balfour in Elasmobranchs,* in which it persists for a very considerable period of larval development. Concerning it he says: "The solidification of the œsophagus belongs to a class of embryological phenomena which are curious rather than interesting, and are mainly worth recording from the possibility of their turning out to have some unsuspected morphological bearings." The condition is so striking a feature in tadpoles of the age we have mentioned, that it can hardly have escaped the notice of other investigators. We have, however, failed to find any mention of it in the works we have consulted. As to any possible morphological bearing, we have nothing to suggest. It is perhaps worth pointing out that it is not concerned with the formation of the lungs, for these we have already described as arising at a much earlier stage, when the œsophagus has a considerable lumen. The explanation may possibly prove to be a physiological rather than a morphological one.

---

* Balfour, "Elasmobranch Fishes," 1878, pp. 217, 218.

The lungs themselves have increased in size, and form a pair of hollow sacs with thick walls, extending back alongside the œsophagus to about the middle of the length of the head kidneys.

3. The Heart and Pericardial Cavity.

Concerning the pericardial cavity at this stage, we have nothing special to say. It communicates freely with the cœlom both dorsally and ventrally.

The heart is more completely twisted on itself than before, and the several divisions more distinctly marked off from one another.

The sinus venosus (Fig. 12, H S) is a wide transverse vessel running across the hinder and dorsal part of the pericardial cavity; it receives at its outer ends the Cuvierian veins, which are now of great size (Figs. 11 and 12, V D); and into its posterior wall open the hepatic veins (V H). In the middle of the anterior wall of the sinus venosus is a large round aperture leading into the auricle.

The auricle (Figs. 11 and 12, H A) is a large, thin-walled, globular sac, lying in the dorsal part of the pericardial cavity, slightly to the right side: there is as yet no trace of the interauricular septum. The auriculo-ventricular aperture is wide, and lies almost horizontally, its actual position, however, varying according to the condition of contraction of the heart.

The ventricle lies below and to the left side of the auricle. Its wall is no longer of uniform thickness, but is strengthened by a reticulum of muscular fibres formed in its substance. These fibres are best developed in the posterior and ventral region, which will afterwards form the apex of the ventricle.

The truncus arteriosus arises from the ventricle at the right hand dorsal corner of its anterior face, as shown in Fig. 11, H M; it then runs forwards, upwards, and slightly to the left, reaching the anterior wall of the pericardial cavity in the median plane (Fig. 12). The walls of the truncus arteriosus are still thin, and its aperture of communication with the ventricle is a narrow one.

In front of the pericardial cavity the truncus divides at once, as before, into right and left branches (Fig. 12), which run outwards and slightly backwards. Each branch quickly divides into two, of which the anterior is the afferent vessel for the first branchial arch and its external gill, while the posterior, after a short course backwards and outwards, divides into the afferent vessels for the second and third branchial arches and their external gills (Fig. 12).

The blood corpuscles have by this time lost almost all their yolk granules, and have the flattened oval shape characteristic of those of the adult frog.

4. THE BLOOD VESSELS OF THE VISCERAL ARCHES.

a. The Mandibular Arch.—The only vessel that can be referred to this arch is what we have called the mandibular artery (Fig. 11, E M), which has relations much the same as in the last stage. It arises from the dorsal aorta, at a level a little posterior to the infundibulum and to the anterior end of the notochord, and runs at first outwards, then downwards and slightly forwards. It lies distinctly in the mandibular arch, close to the anterior surface of the hyomandibular gill pouch. At a level about midway between the dorsal and ventral walls of the pharynx the mandibular artery suddenly enlarges, or rather opens into a wide vein (Fig. 11, V M), which continues its course to the ventral wall of the pharynx, where it communicates by two or more branches with the venous spaces of the floor of the mouth in front of the sucker. A comparison of Figs. 7 and 11 will show that the mandibular vessels have practically identical relations in the two stages.

b. The Hyoid Arch.—The diverticulum from the dorsal aorta, which represents the efferent vessel of the hyoid arch at $6\frac{1}{2}$ mm. (Fig. 7, E H), gradually diminishes in size as the tadpole gets older; at $8\frac{1}{2}$ mm. it can still be recognised as an extremely small bud on the aorta, opposite the middle of the hyoid arch; at 9 mm. it has entirely disappeared (Fig. 11).

At this stage the sole vessel which can be said to belong to the hyoid arch, and this only by comparison with the earlier stages, is the hyoidean vein (Fig. 11, V Y), which lies in the ventral part of the hyoid arch, and communicates below with the venous spaces of the floor of the mouth. Its general relations are the same as at $6\frac{1}{2}$ mm., but it is rather smaller, and more irregular in shape.

c. The First Branchial Arch.—As might be expected from the size of the gill, the vessels of this arch are well developed.

The afferent vessel, or branch of the truncus arteriosus (Figs. 11 and 12, A F.1), is larger than before. On entering the gill it dilates suddenly, and then divides, sending afferent branches to each of the lobes of the gill; these run, as already noticed, along the posterior borders of the several lobes, give smaller branches to the minor lobes, and communicate at the ends of the lobes directly with the efferent vessels (Fig. 12).

The proximal part of the afferent vessel, just before it reaches the external gill, bears along its ventral border a series of small sacculations, usually three or four in number (Fig. 11, A F.1): these lie very close to the surface epithelium, which is slightly raised opposite them, and form the first indication of the internal gills. The earliest age at which we have seen these is in tadpoles of $8\frac{1}{2}$ mm. length. The internal gills are situated ventrally to the external gills and in a line with them, external and internal gills forming apparently one continuous series of structures.

The efferent branchial vessels are in direct continuity, as noticed above, with the afferent vessels at the extremities of the lobes of the gill. At the base of the gill they join the main efferent vessel of the arch (Fig. 11, E F.1), which runs upwards to the dorsal aorta.

Ventral to the gill, this efferent vessel is prolonged downwards, lying close to the anterior border of the afferent vessel. It narrows rapidly and soon loses its lumen, but is continued as a solid cord of cells to the ventral surface of the pharynx, not extending, however, to the bifurcation of the truncus arteriosus (*cf.* Figs. 11 and 12, E F.1). Along this part of their course the afferent and efferent vessels lie very close together, but do not communicate.

The lingual artery is first recognisable about this stage. Though it has as yet no communication with the vessels of the first branchial arch, it may conveniently be described here, as a connection with the first branchial efferent vessel is established very shortly afterwards. At 9 mm. the condition of the lingual artery is shown in Figs. 11 and 12, A L). It consists of a swollen part, which lies immediately in front of the bifurcation of the truncus arteriosus, and from which two vessels arise. Of these, one, the lingual artery proper, runs directly forwards in the floor of the mouth for a short distance, giving off near its base a ventrally directed thyroid artery to the thyroid body (Fig. 11, A R). The other branch runs outwards and backwards, parallel to the afferent vessel of the first branchial arch, and ends blindly a short distance from the blind end of the efferent vessel, and in a direct line with this.

We shall consider the accounts given by other investigators concerning the lingual artery at the conclusion of our description of the next stage, 12 mm. Here we may mention that we have paid very special attention to its early development, and are quite satisfied that at 9 mm. its condition is as described and figured here: at this period it has no communication with any other vessels. Furthermore, we have failed to detect it at any earlier stage, except an immediately preceding one; and are satisfied that in *Rana temporaria* it does not arise from any part of the vessels which we have described in the earlier stages as belonging to the hyoid and mandibular arches. The distance to which the lingual artery can be traced forwards in the floor of the mouth varies in different specimens, but this variation we believe to be connected with the rapid growth forward of the artery after its first appearance.

In Fig. 11, the lingual artery, A L, appears to have very close relations with the hyoid and mandibular veins, V Y and V M. There is really a considerable interval between them, the artery being much nearer the median plane than the veins. Part of the hyoidean vein is represented as cut away in order to show the dilated part of the lingual artery, which would otherwise be hidden by it.

d. The Second Branchial Arch.—The vessels of this arch are practically identical with those of the first arch, and need not be described in detail. The course and relations of the several vessels are the same, and the vessels are of equal size, or if anything slightly larger than those of the first arch. The afferent vessel bears, ventral to the gill, a series of small sacculations (Fig. 11, A F.2) forming the rudiments of the internal gills, which are similar to but rather more prominent than those of the first afferent vessel.

e. The Third Branchial Arch.—The afferent vessel (Figs. 11 and 12, A F.3) is of good size, though smaller than those of the first and second branchial arches. It runs outwards to the small external gill, dilates rather suddenly as it enters it, and then runs along the posterior border of the gill to its tip, where it communicates directly with the efferent vessel. It gives off loops to the lobes of the gill, similar to those of the anterior arches.

The efferent vessel (Figs. 11 and 12, **E F.3**) runs along the anterior border of the gill to its base, and then turns up to the dorsal aorta; below the gill it is continued ventralwards a short distance, ending blindly.

f. The Fourth Branchial Arch.—The vessels of this arch are in almost exactly the same condition as at the last stage, except that the pulmonary artery is rather larger than before. The third and fourth branchial efferent vessels join the dorsal aorta so close together that they appear almost to unite before reaching it.

At this stage pulmonary veins are present as irregular spaces along the inner surfaces of the lungs, in the solid mesoblast connecting the lungs with the œsophagus. These pulmonary veins do not reach the heart, and there is, as we have seen, no trace of a division of the auricle as yet.

g. The Dorsal Aorta.—The relations of the dorsal aortæ are almost precisely the same as at $6\frac{1}{2}$ mm. The course of the anterior arteries is shown in Fig. 11, and the only additional vessels that require mention are the basilar arteries (Fig. 11, **A B**), which arise from the outer ends of the posterior or dorsal infundibular commissure, and run back along the ventral surface of the brain and spinal cord, not far from the median plane.

h. The Veins.—We have nothing special to notice about the veins. The posterior vena cava is commencing to form, but will be more conveniently described at the next stage. Both anterior and posterior cardinal veins are of large size, the anterior cardinal being formed by the union, behind the ear, of a jugular vein returning blood from the brain and the dorsal part of the head, and a facial vein which lies superficially along the side of the head, ventral to both eye and ear.

5. GENERAL CONSIDERATIONS.

These, as already noticed, will be postponed till after the next stage (12 mm.) has been described. The most interesting points arising in connection with the blood vessels at the present stage are the almost complete disappearance of the vessels belonging to the hyoid arch, the formation and relations of the lingual artery, and the first commencements of the internal gills. Of general interest are the perforation of the mouth, the opening out of the gill clefts, and the solid condition of the œsophagus.

VI. THE CONDITION OF THE HEART AND BLOOD VESSELS IN 12 MM. TADPOLES.

1. External Characters.

The most marked difference in external appearance, as compared with the 9 mm. tadpole, is the disappearance of the neck, owing to growth back of the opercular fold, and fusion of its posterior edge with the body wall. The head and body now together form an ovoidal mass about one-third of the entire length of the animal, for the tail has increased considerably in length. The tail is also much wider than before, owing to increase in the height of the ventral and dorsal fins, the latter of which now extends forwards along the head as a low median ridge almost to its anterior end.

The parts about the mouth-opening have grown forwards considerably, so that the nostrils are situated further back on the sides of the head. The eyes, though still small, are more prominent than before. The mouth-opening is bordered by prominent lips, bearing rows of small horny teeth, the lower lip being much the larger. Within the lips lie the large horny jaws.

The two halves of the sucker have now separated from each other, and are placed some distance apart on the under surface of the head.

The opercular membrane is fused at its posterior edge with the body wall, along the right side and across the ventral surface; on the left side of the body it is free, and prolonged backwards as a short tubular spout, through which the tips of the external gills commonly protrude. The opercular membrane is thin and semi-transparent, and through it the gills and heart can readily be seen in a living tadpole.

The abdominal region is smaller than before, and somewhat pyriform in shape, tapering posteriorly to the long conical rectal spout, which opens at the margin of the ventral fin.

At the base of the rectal spout, in the angle between this and the muscles of the tail, the posterior limbs are visible as a pair of small rounded papillæ.

In the tail the division of the muscles into segments is very apparent, the myotomes, as in the earlier stages, being > shaped, with the angle directed forwards, and the dorsal limb of the > shorter than the ventral, as in the adult Amphioxus.

2. Internal Anatomy.

The brain, which has advanced considerably in development, the cerebral hemispheres being now well formed, stops a considerable distance from the anterior end of the head, this change being due to the large growth forwards of the labial region (Fig. 13).

The mouth leads into a buccal cavity, which is rather small, and squarish in transverse section. Both its roof and floor bear prominent papillæ, but there is no trace of the tongue. The posterior nares now open into the roof of the buccal cavity by apertures guarded by valve-like folds of epithelium.

The pharynx, with which the buccal cavity is continuous posteriorly, is shallow dorso-ventrally, but very wide from side to side. In horizontal section it is triangular, with the apex in front; the posterior wall being almost straight, and perforated dorsally in the median plane by the very narrow œsophagus.

The gill clefts are slit-like, and perforate the sides and ventral wall of the pharynx. As seen from the ventral surface, the slits run outwards and backwards obliquely across the floor of the pharynx, so that a single transverse section may cut all four branchial arches and clefts of both sides, the anterior arches appearing at the outer sides of the section, and the posterior arches next to the median plane, (*cf.* Fig. 15, in which these relations are shown at a rather later stage).

With regard to the individual clefts, the hyomandibular cleft does not open to the exterior at this, or, so far as our observations go, at any other period. The first branchial cleft opens, but by a very small aperture, into the anterior end of the opercular cavity. The second, third, and fourth branchial clefts open widely into the opercular cavity.

The operculum, or opercular membrane, arises, as already noticed (Fig. 12, O F), as a backwardly projecting fold from the outer and posterior margin of the hyoid arch. The folds commence at the sides, but very early extend down to the mid-ventral line, so as to form one continuous flap, which grows back as a kind of hood over the sides and ventral surface of the throat, boxing in the gills. The free posterior edge of the operculum soon fuses with the body along the right side, about the level of the swelling caused by the head kidney, a slight fold from the skin of the neck rising up to meet it; the fusion rapidly extends across the ventral surface towards the left side, where the spout-like opening is formed, through which the water is ejected from the opercular cavity in the act of expiration.

Along the inner or pharyngeal borders of the branchial arches, series of small transverse folds arise, which project into the cavity of the pharynx and form the rudiments of the filtering processes, which in later stages attain remarkable complexity (cf. Fig. 15, F. 1, 2, 3, 4).

Across the floor of the pharynx run two horizontal folds, which meet each other posteriorly. These folds run from the sides of the pharynx obliquely inwards and backwards, meeting in the median plane just in front of the glottis; their anterior edges are attached to the floor of the pharynx, while their posterior edges project freely backwards, overlapping the ventral ends of the gill arches; the purpose of these velar folds, as they may be called, appears to be to ensure that the water taken into the mouth for respiration is distributed equally to all the gill clefts.

With regard to the gills; the first branchial arch bears on the left side a well-developed external gill, which commonly projects through the opercular spout. On the right side the external gill is also present, but is very small and shrunken. On both right and left sides the first branchial arch bears along its ventral border a double row of internal gill tufts, hanging down into the opercular cavity (cf. Fig. 15, G L.1).

The second branchial arch is very similar; an external gill is present on each side, that of the right side being much the smaller; a double row of branching gill tufts is borne along the ventral border of each arch.

The third branchial arch has a very small external gill, and also a double row of internal gill tufts similar to those of the first and second arches, but rather smaller.

The fourth branchial arch has no external gill, but supports along its ventral border a single row of internal gill tufts.

The anterior limbs are present as small solid rounded processes arising from the body wall opposite the head kidney, and projecting into the opercular cavity close to its dorsal angle. They are at this stage about the same size as the rudiments of the posterior limbs, or very slightly smaller.

The lungs are much larger than before, their actual size, however, varying very greatly in different specimens; in some cases they extend almost to the hinder end of the body cavity, along the dorsal surface of which they lie. The two lungs open in front into a median laryngeal chamber, which communicates with the pharynx by a slit-like glottis, strengthened laterally by cartilaginous bars.

The thyroid body has lost its connection with the floor of the mouth, and now forms a group of pigmented cells lying close to the anterior wall of the pericardial cavity, and nearly divided into right and left halves by a downwardly projecting process of the basihyal cartilage.

The œsophagus is now open, though its lumen is narrow, and in some specimens very irregular. The precise time and the mode in which the lumen is re-established present some curious and interesting modifications, which we are unable to deal with fully here. In nearly all 10 mm. specimens we find the œsophagus solid along part of its length; at 11 mm. its condition varies very greatly in different specimens; and at 12 mm. the lumen is always present. The lungs re-acquire their opening into the pharynx quite independently of the œsophagus, and at a slightly later period. The head kidneys are still very large, and the Wolffian bodies are just beginning to form.

3. The Heart and Pericardial Cavity.

The pericardial cavity communicates with the cœlom by a pair of apertures dorsal to the Cuvierian veins; *i.e.*, the posterior wall of the pericardium is complete dorsally in its median part, but is perforated laterally by the apertures in question, which are the same as the dorsal apertures of communication described in earlier stages.

In the heart, the principal changes are the more marked separation of the several cavities from one another; the formation of the auricular septum, dividing its cavity into right and left auricles; the structural modifications in the wall of the ventricle, whereby it acquires its characteristic spongy structure; and the formation of definite valves in the truncus arteriosus.

The sinus venosus has undergone little change except that the posterior vena cava is now present, and joins the hepatic vein just as this enters the sinus venosus. The opening from the sinus venosus to the auricular portion of the heart, which before was in the median plane, now lies distinctly to the right side, and leads into the right auricle (Figs. 13 and 14).

The auricle is, as already noticed, almost completely divided into right and left cavities. This division is effected by the growth downwards, from the dorsal wall, of a septum with a free ventral edge. This septum appears first, in tadpoles of about $9\frac{1}{2}$ mm. length, as an inwardly projecting ridge running obliquely across the dorsal wall of the auricle from the left anterior to the right posterior wall; the ridge consisting of a band-like thickening of the outer or muscular wall of the auricle, carrying a fold of endothelium before it. The septum, which from the first is more prominent in front, grows rapidly, hanging down into the auricular cavity as a fold with a free ventral edge; at 12 mm. it extends almost to the auriculo-ventricular aperture. The two chambers into which the auricle is thus divided are at first of very unequal size, the left being much the smaller. Owing to the obliquity of the septum the left auricle lies somewhat dorsally to the right, and does not extend so far forward as this.

The pulmonary veins, which are of small size (Fig. 14, V P), run along the inner sides of the lungs, as in the adult. In front they unite together, and the median vein so formed runs forwards over the sinus venosus to open into the dorsal surface of the left auricle.

The outer wall of the ventricle is no thicker than that of the auricle, but the ventricular cavity is much subdivided by ingrowing muscular trabeculæ, many of which are thicker than the ventricular wall itself. These trabeculæ are arranged in a generally radiate manner; they branch and unite together so as to form a sponge-work, in the meshes of which lie blood corpuscles.

The opening from the ventricle to the truncus arteriosus is circular; its margin is thickened, but there are as yet no distinct valves.

The truncus arteriosus is divided internally into proximal and distal portions by a pair of valve-like folds situated at the constricted part of the truncus just before it divides into right and left branches; these folds are equal in size, and placed opposite each other, projecting into the cavity of the truncus from the right and left sides respectively; the aperture between them, leading to the distal part of the truncus, is very small.

The proximal part of the truncus arteriosus, or pylangium, is partially divided by a longitudinal fold projecting into its cavity. This longitudinal fold, which becomes the longitudinal valve of the adult, arises below from the left side of the ventral lip of the ventricular aperture, and runs up the truncus in a spiral manner, passing round to the posterior or dorsal wall and then to the right side. Its upper or distal end is in very close relation with the right valve of the pair described above as marking the separation between the proximal and distal parts of the truncus, and at a slightly later stage it fuses with this right valve.

The longitudinal valve appears first in tadpoles of about 11 mm. length.

Beyond the paired valves the truncus arteriosus divides almost at once into right and left branches (Fig. 14). Each of these runs directly outwards, and speedily divides into three branches, of which the two anterior ones become the afferent vessels for the first and second branchial arches, while the third runs backwards and outwards, and then divides into the afferent vessels for the third and fourth branchial arches.

It may be noted that although valves are present in the truncus arteriosus, the two primary branches of the truncus are still undivided, so that it is difficult to imagine that the valves play any part in regulating the passage of the blood along the several afferent vessels. It is, however, significant that these valves should appear just at the time when the auricular septum is forming, and the lungs are coming into use.

4. THE BLOOD VESSELS OF THE VISCERAL ARCHES (Figs. 13 and 14).

a. The Mandibular Arch.—As in the earlier stages, there is no trace of an afferent vessel belonging to this arch. The diverticulum of the dorsal aorta, which we have described in the previous stage as the mandibular artery (Fig. 11, E M), is readily recognised at 12 mm., though its relations have altered considerably (Fig. 13, A Y). The proximal part, in connection with the aorta, remains the same as before; but while at 9 mm. the mandibular artery opened distally into the large mandibular vein, and so communicated with the system of sinuses in the floor of the mouth above the sucker (Fig. 11, V M), at the stage we are now considering, 12 mm., the mandibular artery has no connection with lacunar spaces, but continues its course as a vessel of small size with distinct walls, forwards, outwards, and downwards in the roof and sides of the pharynx, to the articulation between the cartilage of the hyoid arch and the skull, where it ends in muscles in the neighbourhood of the joint. Before reaching the articulation it gives off an anteriorly directed branch (Fig. 13, A S), which runs forwards and outwards in the roof of the pharynx.

The artery, A Y, Fig. 13, has the course and distribution of the pharyngeal artery of the adult, to which we believe it to correspond. The anterior branch, A S, is probably the posterior palatine artery* of the adult. These arteries acquire new interest if our determination of the pharyngeal artery as formed, at any rate in part, from the efferent vessel of the mandibular arch, proves correct.

The lacunar system of veins in the floor of the mouth, to which we have referred at the earlier stages, diminishes markedly on the degeneration of the sucker, which is already well advanced at 12 mm.

b. The Hyoid Arch.—At the stage now being considered (12 mm.) there is no trace left of the vessels belonging to the hyoid arch.

c. The First Branchial Arch.—The afferent vessel of the first branchial arch (Figs. 13 and 14, A F.1) runs at first almost horizontally outwards and slightly downwards along the ventral border of the arch, lying ventral to the cartilaginous bar of the arch, and behind the efferent vessel; it diminishes in size as it passes outwards and upwards, and ends on reaching the middle of the side wall of the pharynx; it gives numerous branches to the gill tufts, and before entering the arch a small branch to the velar fold on the floor of the pharynx. The afferent vessel of the left side is distinctly larger than that of the right side, apparently in connection with the larger size of the left external gill.

The efferent vessel (Figs. 13 and 14, E F.1) lies just in front of the

---

* *cf.* Ecker, "The Anatomy of the Frog," edited by G. Haslam. Oxford, 1889; p. 224.

afferent vessel, and is continued beyond it, in the dorsal wall of the pharynx, forwards and inwards towards the dorsal aorta; its most dorsal portion turns slightly backwards and opens into the dorsal aorta. It receives numerous efferent vessels from the internal gills, and opposite the dorsal end of the afferent vessel a small vessel returning blood from the external gill.

The ventral end of the efferent vessel is now directly continuous with the lingual artery; the difference between this stage and the preceding one being clearly shown by a comparison of Figs. 13 and 14 with Figs. 11 and 12. The bulb-like enlargement of the lingual artery, that was noticed at the earlier stage (Fig. 12, A L), is still present, and it is by growth of the posterior branch of the artery towards the end of the efferent vessel, E F.1, that the two vessels meet and become continuous with each other. In front of the swelling, the lingual artery runs forwards and then inwards along the floor of the mouth, ventral to the first branchial and hyoidean cartilages, and then onwards to the lower jaw (Figs. 13 and 14, A L).

In some, but not in all, specimens of this age, there is a direct communication between the afferent and efferent vessels of the first branchial arch. This communication, which is shown in Figs. 13 and 14, is effected between the base of the afferent vessel and the part of the efferent vessel immediately dorsal to the bulb-like swelling on the lingual artery. At this point the afferent and efferent vessels lie close together, but separated by a thin plate of deeply staining cells, apparently epithelial in nature. Through this plate, the thickness of which is rather less than the diameter of the vessels at this part, a narrow and somewhat irregular channel runs, which connects the vessels directly together. We have found it extremely difficult to determine the exact time and mode of establishment of this connection. In some specimens it is effected as just described by a narrow and very short channel; in others the two vessels lie very close together, and a hole in the septum between them places them in direct communication with each other; while in others of the same size, and apparently at the same stage of development, we have failed to find any communication at all.

d. The Second Branchial Arch.—Excepting the absence of any vessel corresponding to the lingual artery, the vessels of the second branchial arch agree very closely with those of the first. The afferent vessel (Figs. 13 and 14, A F.2) is of large size, if anything slightly larger than that of the first arch; just before reaching the gills it gives off a small branch to the filters.

The efferent vessel (Figs. 13 and 14, E F.2) lies just in front of the afferent vessel, receives branches from the gills, both internal and external, and then continues its course to the dorsal aorta.

The ventral end of the efferent vessel lies very close to the afferent vessel, and in some specimens a direct communication takes place between the two, in the same manner as in the first arch. As regards the afferent vessel, this communication is effected beyond the origin of the artery to the filters, but ventral to the lowermost gill vessels. The communication, when present, is effected, as in some specimens of the first branchial arch, by a narrow capillary passage perforating a thin plate of deeply staining cells wedged in between the afferent and efferent vessels, the thickness of the plate being about half the diameter of the vessels.

e. The Third Branchial Arch.—The vessels are very similar to those of the second arch, though of smaller size. Their course and relations are shown in Figs. 13 and 14, A F.3 and E F.3. There is a small external gill present, lying dorsal to the internal gills, and having both afferent and efferent vessels. The ventral end of the efferent vessel communicates at its base with the afferent vessel, in the same manner as in the second arch. In our specimens this connection is easier to see, and apparently more constant in the third than in the second arch.

f. The Fourth Branchial Arch.—The vessels of this arch are much smaller than those of the anterior arches. The afferent vessel (Figs. 13 and 14, A F.4) arises as a branch from the afferent vessel of the third arch, some distance from the division of the truncus arteriosus; it gives afferent branches to the gills of the arch.

The efferent vessel receives the efferent gill capillaries, and then runs nearly vertically upwards in the walls of the pharynx, its course being much straighter than that of the more anterior vessels, and nearly the whole of its length being shown in a single sagittal section. It opens into the dorsal aorta immediately behind the opening of the third branchial efferent vessel (Fig. 13); the two vessels sometimes appearing to unite just before reaching the aorta.

From the dorsal part of the efferent vessel, just before it reaches the aorta, and from its posterior border, the pulmonary and cutaneous arteries arise (Figs. 13 and 14, A P, A U); the roots of the two vessels being extremely close together, but really independent. The pulmonary artery runs at first downwards and backwards, and then directly backwards along the outer surface of the lung. The cutaneous artery runs upwards and outwards immediately behind the auditory capsule, between this and the head kidney, and divides distally near the surface of the body into anterior and posterior branches, supplying the dorsal surface of the head and body.

The afferent and efferent vessels of the fourth arch are in direct communication with each other, ventral to the gills, in the same manner as in the anterior arches. We have found this communication generally though not constantly present in specimens of this age.

g. The Dorsal Aorta.—The dorsal aortæ have undergone but little change since the stage last described. The openings of the efferent vessels of the four branchial arches have been described, and also the origin of the mandibular or pharyngeal artery. In front of this latter the aorta on each side runs inwards rather sharply towards the median plane, and then enters the skull, giving off just before doing so the anterior palatine artery, which runs forward in the mucous membrane of the roof of the mouth as far as the nose. On entering the skull as the internal carotid artery it divides into anterior and posterior cerebral arteries, of which the former runs forwards along the side of the brain, while the latter runs backwards and slightly upwards along the side of the infundibulum, and then backwards along the ventral surface of the brain as the basilar artery (Fig. 13, A B). The arteries of the two sides are connected as before by anterior and posterior infundibular commissures, the anterior or ventral commissure being close to the division of the internal carotid into anterior and posterior cerebral arteries, and the posterior commissure lying in the angle between the infundibulum and the floor of the hind brain.

Behind the branchial arches the two aortæ converge and unite about the level of the hinder end of the head kidneys; just before their union each aorta is in connection with a glomerulus, which is slightly smaller than in the earlier stages.

From each dorsal aorta, close to the openings of the third and fourth branchial efferent vessels, and a little in front of these, the occipito-vertebral artery arises and runs upwards round the hinder border of the auditory capsule, passing between the glossopharyngeal and pneumogastric nerves.

Another small branch arises from each aorta just behind the opening of the fourth branchial efferent vessel; this runs upwards and inwards, and joins the basilar artery of the same side. This is apparently the " vertebral artery " of Goette.*

h. The Veins.—We have nothing special to note concerning the veins, except that they are rapidly acquiring definite walls and more uniform calibre, and are losing the lacunar character that distinguished them during the earlier stages. The posterior vena cava is well established as a large median vein lying between the two Wolffian bodies, ventral to the aorta. In front of the Wolffian bodies it leaves the dorsal body wall, and runs forwards and downwards, lying in a deep groove along the left side of the liver, and joining with the hepatic veins just before reaching the sinus venosus.

5. General Considerations.

There are two points that require more detailed consideration at this stage; firstly, the changes undergone by the vessels of the mandibular and hyoid arches, with which the lingual artery may conveniently be taken; and, secondly, the establishment of direct communications between the afferent and efferent vessels of each branchial arch at the base of the gills.

Concerning the vessels of the mandibular and hyoid arches, we may briefly summarise our conclusions as follows :—In each of the four branchial arches, the vessels peculiar to the arch have in *Rana temporaria* a threefold origin : (1) A pair of lacunar spaces, appearing opposite the level of the external gills, and extending ventrally and dorsally ; the larger space is the first to be formed, it lies in the middle and anterior part of the arch, and it becomes the efferent vessel of the gills of the arch, both external and internal ; the smaller space appears later, lies in the posterior part of the arch, extends ventrally but not dorsally, gives rise to the afferent vessel of the gills, and acquires connection through the gill capillaries with

---

* Goette, A., "Die Entwickelungsgeschichte der Unke," Leipzig, 1875, plate xxii., Fig. 377 *a.v.*

the efferent vessel. (2) An outgrowth from the dorsal aorta. (3) An outgrowth from the truncus arteriosus.

These three factors develope in the order here given; they arise quite independently of each other, and have precisely similar relations in all four branchial arches. Later on, the outgrowth from the aorta joins with the dorsal end of the efferent gill vessel, and the outgrowth from the truncus arteriosus joins with the ventral end of the afferent gill vessel, and in this way the gill circulation is definitely established.

In the hyoid arch the same three factors are present, arising in the same order and in the same manner.

At $4\frac{1}{2}$ mm. (Fig. 4) there is in the hyoid arch an elongated lacunar vessel extending almost the entire length of the arch; there is also a small diverticulum of the dorsal aorta (Fig. 4, E H) lying opposite and close to the dorsal end of the venous lacuna, but not meeting it; and there is as yet no outgrowth from the truncus arteriosus.

At 5 mm. (Fig. 6) the lacunar vessel has divided into dorsal and ventral portions, of which the dorsal is blind, while the ventral (Fig. 6, V Y) opens into a lacunar space above the sucker. The diverticulum from the aorta, E H, is larger than before, but is still blind; and there is a small blind diverticulum from the truncus arteriosus, A H.

At $6\frac{1}{2}$ mm. (Fig. 7) the dorsal portion of the lacunar vessel has disappeared; the ventral part, V Y, has the same relations as before; the diverticulum from the aorta, E H, is present, but very small; and the diverticulum from the truncus arteriosus has disappeared.

At 9 mm. (Fig. 11) the diverticulum of the aorta has disappeared, and the only vessel remaining in the arch is the ventral portion of the lacunar vessel, V Y, which has the same relations as before.

At 12 mm. (Fig. 13) the lacunar vessels of the floor of the mouth have become modified into definitely arranged veins, and it is no longer possible to refer any of these distinctly to the hyoidean vessels of the earlier stages.

These stages appear to us to form a consistent history of the changes undergone by the hyoidean vessels, and to enable a satisfactory comparison to be made between them and the vessels of the branchial arches. The only vessel of the typical branchial set not represented in the hyoid arch is the afferent branchial vessel, and in the entire absence of any rudiment of a gill or gill process this can hardly be wondered at. The development of the vessels of the hyoid arch in the frog appears to us to indicate clearly the former ancestral presence of gills on the arch, while the very early stage at which the vessels undergo retrograde changes suggests that this ancestral condition was a very remote one.

Concerning the mandibular arch, our conclusions are as follows:—

At $4\frac{1}{2}$ mm. (Fig. 4) there are no vessels in the mandibular arch.

At 5 mm. (Fig. 6) there is in the lower or ventral portion of the arch, forming the floor of the anterior part of the mouth, a lacunar space, V M, which is in communication with the lacunar spaces into which the hyoidean vein, V Y, also opens. From the dorsal aorta a small diverticulum, E M, arises, which we have no doubt is to be referred to the mandibular arch.

At $6\frac{1}{2}$ mm. (Fig. 7) the two factors previously present have grown considerably, and have met and opened into each other. We now have a diverticulum from the dorsal aorta, E M, of some length, which runs down the mandibular arch and opens into the dorsal end of the mandibular vein, V M. This communication, which is a very definite and unmistakeable one, affords strong evidence in support of our comparison of the lacunar vessel, V M, with the efferent lacunar vessel of the hyoidean or branchial arches.

The diverticulum from the aorta, E M, lies distinctly in front of the hyomandibular cleft, and must beyond doubt be considered as belonging to the mandibular arch.

At 9 mm. (Fig. 11) there is no change of importance in the vessels of the mandibular arch, except that the communication between the lower end of the mandibular vein and the sinuses of the floor of the mouth is narrower than before.

At 12 mm. (Fig. 13) the mandibular artery, A Y, has the course and relations of the pharyngeal artery of the adult. We have studied the intermediate stages, and have no doubt that the dorsal part of this artery, from its origin from the aorta to its point of division, is the same thing as the vessel, E M, in the 9 mm. stage. We have not, however, determined with certainty the time and mode of formation of the artery we have called posterior palatine (Fig. 13, A S), nor the precise fate of the mandibular vein (Fig. 7, V M), into which the mandibular artery opens at 9 mm. We therefore conclude that in the mandibular arch the efferent branchial vessel and the diverticulum from the aorta are present, but that the afferent branchial vessel and the diverticulum from the truncus arteriosus are not present at any stage; *i.e.*, that the vessels of the mandibular arch are so far comparable to those of a branchial arch as to indicate the ancestral presence of gills, but that the modifications undergone by the mandibular arch are more profound than those of the hyoid arch. A further difference between the mandibular and hyoid arches lies in the fact that while the vessels of the latter disappear completely, the aortic diverticulum of the mandibular arch persists as the pharyngeal artery.

Concerning the lingual artery, we have looked out very carefully for any evidence of its belonging to either the mandibular or hyoidean arches, but have entirely failed to find such.

The earliest date at which we have found the lingual artery is about $8\frac{1}{2}$ mm., when it has relations similar to those shown in a 9 mm. tadpole in Fig. 11, A L, but is of rather smaller size. In the frog it appears therefore to arise late in development, and to be at first independent of all other vessels. While satisfied as to the correctness of this account, we are not disposed to attach much importance to it from a morphological standpoint. The mode in which the lingual artery becomes connected with the ventral end of the first branchial efferent vessel strongly suggests that the actual, ontogenetic, development is a modified one. The bulb-like swelling on the lingual artery, which is present from its first appearance, is a curious feature for which some explanation is wanted.

Turning now to the accounts given by other investigators of the development of the vessels in the mandibular and hyoidean arches of Amphibians, both Goette and Maurer have given somewhat brief descriptions.

Goette* describes in *Bombinator* a small vessel as developing in the hyoid arch subsequently to the branchial vessels, and acquiring a connection dorsally with the dorsal aorta or carotid artery just in front of the ear; while ventrally it joins the first afferent branch of the truncus arteriosus. This vessel therefore forms a complete arch connecting the truncus arteriosus and the aorta directly. It soon,

---

* Goette, A., "Entwickelungsgeschichte der Unke," pp. 756, 757, and plate xxi., Fig. 377.

according to Goette, undergoes changes: the middle portion of its course becomes obliterated; its ventral end remains as the thyroid artery, from which the lingual artery arises; while its dorsal end is described as becoming the root of a new artery, the temporo-maxillary, which runs forwards, outwards, and downwards in the mandibular arch.

Goette's description is very incomplete. It is, however, sufficient to show that his temporo-maxillary artery is what we have described as the pharyngeal artery, and his account agrees with ours in placing this artery in the mandibular arch. The independent origin of this artery from the aorta, he has apparently overlooked. Concerning the hyoid arch, our accounts agree in describing a vessel in the arch which at first does not open into either aorta or truncus (*cf.* Fig. 4, E H' and V Y). The connection with the aorta we find later in *Rana*, but not the connection with the truncus. In the early disappearance of the hyoid vessel along the greater part of its length, the two accounts agree.

Maurer's* observations were made on *Rana esculenta*. He finds in tadpoles of 4 mm. length a vessel which he terms the hyomandibular artery, which arises from the truncus arteriosus, runs forwards and outwards across the hyoid arch into the mandibular arch, and then upwards and inwards, to end blindly above the pharynx, the dorsal aorta not being formed as yet.

In 5 mm. tadpoles Maurer describes the hyomandibular artery as arising not from the truncus arteriosus directly, but from the ventral end of the first branchial efferent vessel. At 6 mm. the relations are much the same, the hyomandibular artery being still connected with the ventral end of the first branchial efferent vessel, and being described as a ventral continuation of the latter, which at a later stage, 9 mm., is said to become the "external carotid" or lingual artery.

We have found it impossible to reconcile this description with our own observations on tadpoles of *Rana temporaria*, and can only suppose that the differences in the mode of development of these vessels in allied species are extraordinarily great, or else that Maurer has been misled by the large venous spaces in the floor of the mouth,

---

* Maurer, "Die Kiemen und ihre Gefässe bei Anuren und Urodelen Amphibien," "Morphologisches Jahrbuch," xiv., pp. 175-222.

which he does not specially mention, and has described a connection between the lingual artery and the vessels of the mandibular arch which does not really exist. That the former alternative is, however, quite possible, is shown by the very different development of the mandibular vessels in the two species. Maurer's description, which on this point leaves no room for doubt, shows that in *Rana esculenta* there is a complete vessel in the mandibular arch at a stage when the dorsal aorta has not yet formed; that the mandibular vessels are in fact among the very earliest to appear. In *Rana temporaria*, on the other hand, we find no trace of a vessel in the mandibular arch until long after the aorta is completed, and at no period do we find this vessel in continuity with the heart.

The general conclusions to be drawn concerning the vessels of the mandibular and hyoidean arches appear to be that in both arches vessels are found which in mode of development and in general relations conform to the type of the vessels in the branchial arches. In *Rana temporaria* the hyoid vessels appear earlier, and correspond more closely to the branchial type than do the mandibular vessels; while in *Rana esculenta*, according to Maurer's description, the reverse appears to be the case.

The establishment of direct communication between the afferent and efferent vessels of the branchial arches requires some further notice. There is no point in our investigations which has given us greater trouble, or about which we have obtained such conflicting evidence from different specimens. It is not always possible to determine the presence of a small hole of communication between two closely adjacent vessels by examination of a series of sections; nor is it possible in every case in which such a hole or passage is detected to make quite certain that it is natural and not artificial. We, therefore, desire to speak on this point with considerable reserve, fully acknowledging that confirmation of our results is needed before their acceptance can be claimed. We have found it a good test to compare the two sides of the body in any doubtful case: an opening which is present on both sides in identical positions may fairly be assumed to be natural.

According to our own observations detailed above, the direct communications between the afferent and efferent branchial vessels, which become established in tadpoles of about 12 mm. length (*cf.*

Fig. 13), place the efferent vessels for the first time in direct connection with the truncus arteriosus.

According to Maurer's description, this is not the case in *Rana esculenta*, for in this species the efferent or primary vessel is at first the only vessel in the arch, and opens directly into the truncus arteriosus. On the formation of the afferent or secondary vessel, the efferent vessel loses its connection with the truncus, regaining it, however, at a stage corresponding to that shown in Fig. 13, and in a manner exactly corresponding to what we have described in *Rana temporaria*. This direct communication between afferent and efferent vessels is of great physiological importance, for as soon as it is established a path is open for the blood from the heart to the aorta without passing through the gills; and it is the gradual enlargement of this direct passage that leads to the atrophy of the gills, and the conversion of the animal from a gill-breathing to a purely air-breathing form.

The aperture or channel of communication between the afferent and efferent vessels of the first branchial arch is of especial interest, as it is in direct connection with it that that problematical structure, the carotid gland, is formed at a later stage.

The development of the carotid gland takes place in the following manner: At 12 mm., as shown in Figs. 13 and 14, the lingual artery and first branchial efferent vessel form one continuous vessel. At the base of the lingual artery is a small bulb-like swelling, which was present at an earlier stage, while the lingual artery was still an independent vessel (Fig. 11). Immediately beyond this bulb the afferent and efferent vessels are in direct communication, the passage being a single narrow one, traversing a small plate of the deeply staining cells described previously.* In the later stages this passage becomes plexiform, there being now three or four openings into the afferent vessel, and about the same number into the efferent, one or more of the latter opening directly into the bulb-like swelling at the base of the lingual artery. This plexiform communication becomes the carotid gland, the history of its formation showing that it is not to be regarded as a persistent or modified part of a gill, but that

---

* According to Maurer this epithelial plate is budded off from the epithelium of the first branchial cleft. *Vid.* "Schilddrüse, Thymus und Kiemenreste der Amphibien," "Morphologisches Jahrbuch," xiii., 1888, p. 321.

it is a specially acquired structure, formed by elaboration of the direct passage between afferent and efferent branchial vessels which is present in a simpler form on the hinder arches as well.

There are two other points that deserve some attention.

In the first place, Goette* figures and describes the third and fourth branchial efferent vessels in Bombinator as not opening directly into the dorsal aorta, but as uniting together at their dorsal ends and running back as the pulmonary artery. Our descriptions and figures, with those of Maurer, show that this is not the case in Rana, the third and fourth branchial efferent vessels opening into the aorta in the same manner as the first and second vessels.

Lastly, our figures and descriptions show clearly that both the cutaneous and pulmonary arteries arise from the fourth branchial efferent vessel, a point in which we agree with Boas† and differ from Goette, who derives the cutaneous artery from the third arch.

VII. THE CONDITION OF THE HEART AND BLOOD VESSELS IN 21 mm. TO 23 mm. TADPOLES.

1. EXTERNAL CHARACTERS.

In tadpoles of from 21 mm. to 23 mm. length the general proportions have not altered much. The head and body form a wide rounded mass, about one-third of the entire length of the animal. The lips bordering the mouth are very prominent and provided with numerous teeth. The opercular spout is a cylindrical tube running from the ventral surface backwards and upwards along the left side of the body, and opening by a rounded aperture about the middle of the length of the body. The suckers have entirely disappeared : the nostrils are small, the eyes larger and more prominent than before.

The rectal spout opens at the ventral border of the fin by an aperture inclined somewhat obliquely towards the right side. At the base of the spout are the hind limbs, which, though larger than before, are still mere papillæ, about half the length of the rectal spout.

---

* Goette, "Die Entwickelungsgeschichte der Unke," p. 753, and plate xxi., Fig. 377.
† Boas, "Ueber den Conus Arteriosus und die Aortenbogen der Amphibien," "Morphologisches Jahrbuch," vii., 1881, p. 548.

Owing to the greatly developed lymph spaces, the body of the tadpole, when viewed by transmitted light, appears surrounded by a semi-transparent margin or fringe.

2. INTERNAL ANATOMY.

The mouth is still on the ventral surface, but is more anteriorly placed than before. The buccal region has grown forward considerably, so that the anterior end of the brain is now a long way behind the front of the head. The pharynx has much the same shape as before; on its floor are two rows of papillæ, converging posteriorly. At the hinder and lateral part of the pharynx are the gill clefts, overhung in front by the velar folds. There is as yet no tongue.

The first branchial cleft opens into the side of the pharynx, just in front of the outer end of the velar fold, and at the hinder end of a groove (cf. Fig. 14, C G), which runs obliquely backwards and outwards across the floor of the mouth.

The filters on the inner borders of the branchial arches are greatly developed: the inner border of each arch is raised into a prominent longitudinal fold, which bears on both its faces closely set transverse ridges, notched along their free edges in a complicated manner into series of tooth-shaped processes (cf. Fig. 15, F.1,2,3,4). These filters are extremely vascular, but as the blood is returned from them to the somatic veins, it is probable that they are not actively respiratory.

The external gills are absent, and the internal gills (Fig. 15, G L. 1-4) are arranged as before, but are rather better developed.

The fore limbs project into the upper angles of the opercular cavity as solid blunt processes, about twice as long as they are broad.

The alimentary canal has greatly increased in length, the intestine being much convoluted. The head kidneys are greatly reduced in size, and their tubules are in process of rapid degeneration. The Wolffian bodies are now well developed; their anterior ends are situated some distance behind the head kidneys, while their posterior and larger ends extend into the hinder part of the body cavity.

The lungs are large, thin walled, and filled with air; they lie in the abdomen close to its dorsal wall, above all the other abdominal viscera and about the level of the notochord.

The thymus forms a very conspicuous body in transverse sections. It consists (Fig. 15, T) of a pair of rounded masses of closely compacted spherical cells, lying in the sides of the roof of the pharynx below the anterior part of the auditory capsules.

One of the most marked features of the present as compared with the earlier stages, is the enormous development of the subcutaneous lymph spaces. These form sacs of literally enormous size (Fig. 15, Y), filled with a coagulable fluid, and almost completely surrounding the sides and ventral surface of the body and head. It is owing to these lymph spaces that spirit specimens of tadpoles in these later stages almost invariably have a shrivelled appearance.

3. The Heart and Pericardial Cavity.

The pericardial cavity has the same communications as before with the cœlom.

In the heart itself there are no great changes. The left auricle is larger relatively than at 12 mm., but is still much the smaller of the two. Owing to the position of the inter-auricular septum, the left auricle lies somewhat obliquely behind the right one.

The auriculo-ventricular aperture is much narrowed by valvular folds. The ventricle (Fig. 15, H V) is in much the same condition as before: its outer wall is if anything rather thinner than that of the auricles, except at the places where the muscular trabeculæ arise from it. These are now thicker and much more numerous than before. They are arranged so as to leave a central cavity, free from trabeculæ, and leading from the auriculo-ventricular aperture to the aperture into the truncus arteriosus. The rest of the cavity of the ventricle has the appearance of a spongy reticulum, owing to the number of the trabeculæ and the intricate way in which they cross. This mode of formation of the musculature of the ventricle by an internal reticular mesh-work, in place of solid thickening of its walls, allows the blood, which occupies all the meshes of the sponge-work, to come into immediate relation with every part of the ventricle, and so explains the absence of nutrient vessels in the ventricular walls, first noticed by Hyrtl.

The truncus arteriosus has thick muscular walls, and contains valves arranged as before, there being as yet no arrangement for regulating the distribution of the blood from the two auricles to the several aortic arches. In front of the pericardial cavity the truncus arteriosus divides at once into right and left branches, each of which is still single at its base. Each branch, after a short course outwards, divides into a smaller anterior branch, which is the afferent vessel for the first branchial arch, and a larger posterior part which again divides almost at once into the second branchial afferent vessel, and a posterior vessel which after a short course outwards and backwards divides into the afferent vessels for the third and fourth branchial arches.

In sections of tadpoles of about this stage the blood vessels are partially filled with blood clots: these are not seen during the earlier stages of development, in which the vessels appear empty but for the corpuscles. These latter have by this time the shape and appearance of those of the adult.

4. THE BLOOD VESSELS OF THE VISCERAL ARCHES.

a. The Mandibular Arch.—The pharyngeal artery, which is now the only vessel referable to this arch, has the same relations as before. It arises from the aorta rather further forwards, just before the latter enters the skull as the internal carotid artery.

b. The Hyoidean Arch has, as described in the last stage, no vessels referable to the branchial series.

c. The First Branchial Arch.—The afferent branch of the truncus arteriosus runs outwards to the base of the first gill: here it enlarges rather suddenly, and gives off branches to the filters on the first branchial arch. It then continues along the arch, giving off numerous branches to the internal gills.

The efferent vessel communicates with the afferent at the base of the arch by a carotid gland, which consists of two or three narrow irregular channels of a spongy appearance running in the septum between the two vessels, and opening into both of them. Beyond the carotid gland the lingual artery runs forwards in the floor of the mouth. In the branchial arch itself the relations of the efferent vessel are the same as at the previous stage.

d. and e. The Second and Third Branchial Arches.—The vessels of these arches have undergone scarcely any change as compared with the 12 mm. stage. In both arches we have found the afferent and efferent vessels to be in direct communication at the base of the gills:

this communication is a very narrow one, and is seen more readily in the third than in the second arch. It is shown on the right side of Fig. 15 in the third arch (Fig. 15, A E).

f. The Fourth Branchial Arch.—We are uncertain as to what is the normal condition of the vessels in this arch at the stage we are describing. In some specimens we find well-formed afferent and efferent vessels as in the earlier stage, while in others, although a gill is present, we have seen only a single vessel. We are inclined to regard the vessels of this arch as in the act of transformation, the afferent and efferent vessels communicating freely at the base of the gill, and the upper or distal part of the afferent having already disappeared in some specimens.

From the dorsal part of the efferent vessel, just before it opens into the aorta, the cutaneous and pulmonary arteries are given off as before.

g. The Dorsal Aorta.—We have nothing to say concerning the aorta at this stage, except to note that the glomerulus is slightly smaller and more pigmented than before. The lung, which lies immediately ventral to the glomerulus on each side, is fused along its outer wall for a short distance with the somatopleure opposite the head kidney. The part of the body cavity in which the glomerulus lies becomes thus partially shut off from the rest of the cœlom, with which, however, it remains in communication posteriorly, and also, though by a more restricted passage, anteriorly, in front of the root of the lung.

5. GENERAL CONSIDERATIONS.

There is nothing special to note at this stage except the formation of the carotid gland, which has already been considered under the previous stage.

VIII. THE CONDITION OF THE HEART AND BLOOD VESSELS IN TAILED FROGS DURING THE METAMORPHOSIS.

1. EXTERNAL CHARACTERS.

The stage we have selected for description is one in which the total length is 22 mm., of which 14 mm. belong to the head and trunk, and 8 mm. to the tail.

The limbs are well developed; the hind limbs when fully extended reaching a little beyond the end of the tail; the toes are well formed and webbed. The fore limb of the left side projects through the opercular spout; that of the right side is protruded through the opercular fold, in which it forms a ragged hole. In slightly younger specimens the right fore limb still lies within the opercular cavity.

The mouth has changed its appearance altogether: the horny jaws and frilled lips of the tadpole have gone, and the mouth aperture is now slit-like and shaped as in the adult frog, though not extending so far back.

2. INTERNAL ANATOMY.

A well-developed tongue is present, attached to the floor of the anterior part of the mouth, and with its tip bifid and directed backwards as in the adult.

The gill arches and gills are still present, and are arranged very much as in the earlier stage. Velar folds of the floor of the pharynx are present, overlapping the gills, but are somewhat smaller than before. The filters are exceedingly complicated and well developed, but the gill tufts themselves are, relatively to the bulk of the animal, much smaller than before.

The alimentary canal behind the pharynx has proportions much closer to those of the adult than it had in the earlier stages. The stomach is more dilated than before, and has the adult shape and position: the small intestine has shortened considerably: the liver is large, and the gall bladder, which lies on the right side, behind the right lobe of the liver, is of relatively enormous size.

The head kidneys are very much reduced in size, and are now relatively insignificant bodies, in which, however, the hindermost of the three nephrostomes on each side are still present. The glomerulus is still present, having the same appearance and relations as in the earlier stages, but is of distinctly smaller size. The Wolffian bodies are large, especially at their hinder ends. The reproductive organs, showing as yet no difference of sex, are present as a pair of narrow longitudinal bands lying ventral to the Wolffian bodies, and along the inner borders of their anterior halves. The fat-bodies are present, in close connection with the anterior ends of the reproductive organs. The allantoic bladder forms a saccular diverticulum from the ventral surface of the rectum as in the adult.

The lungs lie close against the dorsal wall of the body, and extend almost its whole length.

The brain has all the parts of the adult, and very nearly the adult proportions. In the skeleton, ossification is advancing rapidly, and the limb girdles and sternum are already well developed.

The two halves of the thyroid body, which have now separated completely from each other, lie in close contact with the anterior wall of the pericardial cavity. Each half is a mass of vesicles of irregular size and shape, lined by a single layer of cubical epithelial cells, and with the cavities of the vesicles filled by a coagulable fluid. The thyroid body receives blood from the thyroid branch of the lingual artery, and returns it by a short vein to the inferior jugular vein.

The subcutaneous lymphatic spaces are enormously developed, especially in the sides and floor of the head.

3. The Heart and Pericardial Cavity.

The pericardial cavity still communicates with the cœlom by a pair of apertures near its dorsal surface, situated below and at the sides of the median laryngeal chamber.

The sinus venosus is much as before. Of the vessels in communication with it the posterior cardinal vein has diminished greatly in size, and a subclavian vein is present returning blood from the fore limb.

The aperture from the sinus venosus to the right auricle is now guarded by a valvular fold. The left auricle has increased in size, but is still a good deal smaller than the right. Both auricles have their walls thickened by muscular strands, which form interlacing reticular ridges on their inner surfaces. Near the outer sides of the auricles some strands cross their cavities as in the ventricle.

The ventricle is relatively larger than before. The internal network of interlacing muscular strands is closer and more intricate than before, but the true outer wall of the ventricle still remains thin, as thin indeed as the wall of the auricles. The muscular reticulum is absent from the centre of the ventricle, where a wide open channel is left, leading from the auricular aperture to that of the truncus arteriosus. The aperture from the ventricle to the truncus arteriosus is guarded by pocket valves.

The truncus arteriosus may as before be divided into a proximal part or pylangium which receives the blood from the ventricle, and a distal part or synangium which divides on each side into the afferent branchial vessels.

The pylangium and synangium are separated from each other by well-formed valves. Of these there were two at the stage last described, in the form of right and left flaps, placed transversely to the vessel. In place of these two simple valves, there are now three pocket valves, with free upwardly-directed edges. These valves are very unequal in size, and consist of a large one on the right side of the truncus, and two smaller ones, anterior and posterior, on the left side. They correspond in the order given to the valves numbered by Boas* 1, 2, and 3 respectively.

Below these valves the cavity of the pylangium is single, but is partially divided by the spiral valve. This valve has the same relations as before. It is a longitudinal inwardly projecting fold of the pylangium, which, commencing at the ventricular aperture, where it lies to the left of the pocket valves, runs up somewhat spirally along the posterior or dorsal wall of the pylangium, and ultimately to its right side, where it ends immediately below the right hand valve, valve No. 1 of Boas, of the three between pylangium and synangium. The condition is therefore practically that of the adult, except that the upper end of the spiral valve, though lying just below valve No. 1, has not yet fused with it.

The distal part of the truncus, or synangium, is now divided internally by a vertical partition into anterior and posterior vessels, of which the anterior divides on each side to form the first and second branchial afferent vessels, while the posterior similarly divides to form the third and fourth branchial afferent vessels. In the synangium itself this division is an absolute one, the first and second arches being shut off completely from the third and fourth. The septum which effects this division ends in a free lower edge immediately above the three pocket valves that separate the synangium from the pylangium.

---

* Boas, "Ueber den Conus Arteriosus und die Aortenbogen der Amphibien," "Morphologisches Jahrbuch," vii., 1881, pp. 502-507, and plate xxiv., Figs. 15 and 16.

The changes required to convert this condition of the truncus arteriosus into that of the adult are very slight. We have not ourselves followed out these changes, and would refer for details of the mode in which they are effected to the paper by Boas quoted above.

4. THE BLOOD VESSELS OF THE VISCERAL ARCHES.

a. and b. The Mandibular and Hyoid Arches.—These have undergone no changes of importance since the last stage: the vessels of these arches have, in fact, already acquired their adult arrangement.

c. The First Branchial Arch.—Both afferent and efferent vessels are present, and the relations are almost exactly those of the last stage, the only change of importance concerning the communication between the afferent and efferent vessels at the base of the gill.

In the former stage this communication was effected by a single short passage of capillary size, while at the present stage it takes place through a well-developed plexiform carotid gland. From this plexus the lingual artery arises by two roots, which embrace the gland, one lying above it and one below.

d. The Second Branchial Arch.—The vessels of this arch are rather larger than those of the first branchial arch; they are also larger on the left side than the right. We have found extreme difficulty in determining whether a direct connection between the afferent and efferent vessels of this arch is or is not present at this stage. In some specimens such a connection appears to exist, while in others we have entirely failed to find it. If it is constantly present, it is certainly far less obvious in this than in the other arches.

e. The Third Branchial Arch.—The vessels of this arch are decidedly smaller than those of the first or second arches. A direct connection between the afferent and efferent vessels is constantly present at the base of the gills.

f. The Fourth Branchial Arch.—Both afferent and efferent vessels are present; they are smaller than those of the other arches, and lie very close together. Direct communication between the two is present in some cases, but we have not been able to detect it in all specimens. The dorsal end of the efferent vessel opens into the aorta immediately behind the third efferent vessel, but independently of this. Shortly before reaching the aorta, the fourth efferent vessel gives off the pulmonary and cutaneous arteries, whose course and relations are as before.

5. General Considerations.

We have been somewhat puzzled at this stage by our failure in some cases to find the direct connections between afferent and efferent vessels which appear to be constantly present at an earlier stage. At present we are unable to explain our results in this respect, unless there should prove to be great individual variability in the precise time and mode of appearance of these communications in the several arches. It is in some cases a matter of extreme difficulty to determine from sections whether such a communication does or does not exist; and in some specimens, whose appearance suggests in the strongest manner that the direct connection is present, we have been unable to trace it out in a satisfactory and conclusive manner. To follow with certainty a single tortuous capillary passage through half-a-dozen or more sections, in which it may be cut in all conceivable planes, is not always possible.

Boas*, who has investigated these later stages by injecting the vessels and then dissecting them out, has found direct connections present between the afferent and efferent vessels of the first and second branchial arches in a tadpole in which the tail had begun to shrink, but was still long: the third and fourth arches were unfortunately destroyed in the specimen. The communications in the first and second arches were so free that the injection mass passed across from the base of the afferent to the efferent vessel, leaving the dorsal part of the afferent vessel empty.

IX. THE CONDITION OF THE VESSELS OF THE BRANCHIAL ARCHES IN YOUNG FROGS TOWARDS THE END OF THE FIRST YEAR.

This is the last stage that we propose to deal with. The frogs, which are now several months old, have lost their gills completely, though remnants may persist in a functionless condition. There is

---

* Boas, "Ueber den Conus Arteriosus und die Aortenbogen der Amphibien," "Morphologisches Jahrbuch," vii., 1881, p. 547.

no object in giving with regard to this stage any details concerning the general anatomy, or the condition of the heart, or the vessels of the anterior part of the head. We therefore pass at once to the consideration of the vessels of the four branchial arches.

In each arch there is now only a single vessel, or aortic arch, as it is commonly termed. This aortic arch is shown by stages intermediate between the present and the last described one, to be formed by enlargement of the direct communication between the afferent and efferent vessels, followed by gradual shrinking and atrophy of the part of the afferent vessel above the aperture of communication. Each aortic arch thus consists of the proximal or cardiac portion of the afferent branchial vessel together with the whole length of the efferent branchial vessel of the arch.

Of the four aortic arches, that of the first branchial arch becomes the carotid arch of the adult. It is interrupted opposite the base of the lingual artery by the carotid gland, which is the persistent plexiform communication between the former afferent and efferent vessels, from which the lingual artery still arises by two roots. The portion of the dorsal aorta between the points of opening of the first and second aortic arches remains an open tubular vessel for some time, and may even retain its lumen in the adult. More usually, however, the cavity becomes obliterated, and the walls of the vessel merely form a pigmented band connecting the dorsal ends of the first and second, i.e., the carotid and systemic arches, together. After the obliteration of this part of the aorta, the blood passing along the carotid arch is distributed exclusively to the head.

The second aortic arch, in the second branchial arch, becomes the systemic arch of the adult, and requires no further description.

The third aortic arch, in the third branchial arch, disappears altogether. In young frogs of the first year it loses its connection with the aorta, and then gradually shortens up, the distal part becoming a solid cellular cord, and the proximal or cardiac part retaining for a time its lumen. Before the end of the first year this vessel has entirely disappeared.

The fourth aortic arch, in the fourth branchial arch, becomes the pulmo-cutaneous arch of the adult. It retains its communication with the aorta for some time after the dorsal part of the third arch has atrophied; but before the end of the first year the part of the fourth arch above the origin of the cutaneous artery loses its cavity and becomes a solid cord, so that the fourth or pulmo-cutaneous arch no longer opens into the aorta.

The condition of the arteries is now that of the adult.

Our description of these final stages in the formation of the arterial system of the adult agrees exactly with the description given by Boas in his paper quoted above, and it is a source of satisfaction to us that we have been able to confirm the results of so careful a worker.

We conclude this paper with a summary of the principal conclusions to which we have been led.

X. EPITOME OF RESULTS.

All our results apply to *Rana temporaria* alone, unless otherwise stated.

A. *Development of the Heart.*

1. The heart developes before any of the vessels of the visceral arches, and before the dorsal aortæ.
2. The muscular wall of the heart is formed from the splanchnic layer of mesoblast. The inner or endothelial wall of the heart is derived directly from the hypoblast of the ventral wall of the pharynx and of the liver diverticulum.
3. The heart is from the first in connection at its posterior end with the veins of the yolk-sac and of the liver.
4. The heart becomes twisted on itself, and its several chambers marked off by constrictions, before its anterior end acquires any connection with blood vessels.
5. The auricle is at first single, but later becomes divided into right and left auricles by a septum growing down from its dorsal wall.
6. The wall of the ventricle remains thin throughout the whole of development. The apparent thickening which occurs is due to the development of an internal muscular reticulum. The absence of nutrient vessels in the wall of the ventricle is explained by this arrangement.
7. The valves of the truncus arteriosus are established before the metamorphosis.

B. *Formation of the Blood Vessels and Blood Corpuscles.*

8. The blood vessels arise as irregular lacunar spaces in the mesoblast, formed by widening of the intervals between adjacent mesoblast cells. These spaces open into each other, and so form continuous channels.
9. The mesoblast cells surrounding these channels become converted into the epithelial walls of the vessels.
10. The further growth of the vessels is sometimes effected by the formation of solid cellular cords, continuing the lines of the vessels. The axial cells of these cords break down, thus giving rise to tubular vessels.
11. There are at first no blood corpuscles. When these appear, they are formed from the walls of the vessels themselves.
12. The blood corpuscles are true cells. They are at first spherical in shape, and laden with yolk granules. By the time the mouth opens, the yolk granules are much diminished in size and number; and shortly afterwards they disappear, and the corpuscles acquire their adult shape and appearance.

C. *Development of the Aorta and of the Vessels of the Visceral Arches.*

13. The first vessels to appear, except the vitelline veins and heart, are the dorsal aortæ. These arise on each side as a number of isolated lacunar spaces along the roof of the pharynx, which open into one another and so form continuous vessels.
14. In the four branchial arches, blood vessels are formed on a definite plan. In the first and second branchial arches, these vessels appear immediately after the dorsal aortæ: in the third and fourth branchial arches they arise later.
15. In each branchial arch there appears a lacunar efferent vessel at the level of the external gill; this extends dorsally, and meets, and opens into a diverticulum of the dorsal aorta. It also extends ventralwards towards the truncus arteriosus, but does not open into this.
16. Immediately behind the efferent vessel, at the level of the gill, an afferent vessel appears; this is also lacunar, and at first independent; it soon acquires connection with

the efferent lacunar vessel by capillary loops in the gill, while ventrally it grows downwards in the arch, and meets and opens into an afferent diverticulum from the truncus arteriosus.

17. The course of the circulation is therefore from the truncus arteriosus to the afferent branchial vessel, then through the capillary loops of the gills to the efferent branchial vessel, and then on to the dorsal aorta.

18. In *Rana esculenta*, according to Maurer, the condition is very different: the efferent vessel is at a very early stage connected with the truncus arteriosus, and affords a direct passage to the aorta. In one single specimen of *Rana temporaria* we have found a similar condition.

19. In the hyoid arch vessels appear at an early stage, which admit of close comparison with the vessels of a branchial arch. They soon, however, undergo degenerative changes, and disappear entirely at an early stage.

20. In the mandibular arch vessels are formed which depart widely from the arrangement in a branchial arch, but yet admit of comparison with this. Of these vessels, the only one that persists is the diverticulum of the dorsal aorta, which becomes the pharyngeal artery of the adult frog.

21. The internal carotid artery is formed as an extension forwards of the dorsal aorta of each side.

22. The lingual artery arises independently in the floor of the mouth: it soon becomes connected with the ventral end of the efferent vessel of the first branchial arch.

23. Close to the ventral ends of the efferent branchial vessels direct connections become established between the afferent and efferent branchial vessels. These direct connections are formed in all four branchial arches: they are ventral to the gills, and afford a direct, but at first very narrow passage, by which the blood can pass from the truncus arteriosus to the efferent vessel, and so to the dorsal aorta, without passing through the vessels of the gills. They appear first in tadpoles of about 12 mm. length.

24. The carotid gland is not a persistent gill, but is formed by elaboration of the direct connection between the afferent and efferent vessels of the first branchial arch, which, in place of being a simple passage, becomes plexiform.
25. The pulmonary and cutaneous arteries develope at an early period as diverticula from the dorsal part of the efferent vessel of the fourth branchial arch, shortly before this joins the aorta.
26. By enlargement of the direct connections between afferent and efferent vessels at the bases of the branchial arches at the time of the metamorphosis, an increasing quantity of blood passes from the heart to the aorta without going through the gills. Increased work is thus thrown on the lungs as respiratory organs, while the gills gradually atrophy.
27. The first branchial efferent vessel becomes the carotid arch of the adult.
28. The second branchial efferent vessel becomes the systemic arch of the adult.
29. The third branchial efferent vessel loses its connection with the aorta, gradually atrophies, and ultimately disappears completely about the end of the first year.
30. The fourth branchial efferent vessel becomes the pulmo-cutaneous arch of the adult: the part of the vessel dorsal to the origins of the pulmonary and cutaneous arteries, which at first opens into the aorta, becomes solid, and then disappears.
31. The part of the aorta between the carotid and systemic arches usually becomes solid, but may remain open even in the adult.
32. The glomerulus of the head kidney arises as a bulging of the ventral surface of the aorta, which soon becomes sacculated. The glomerulus is large so long as the head kidney is active, but then diminishes greatly in size.

D. *General Points.*

33. Throughout the early stages of development there is a striking resemblance in arrangement, relations, and pro-

portions between the arterial and venous systems of the tadpole and those of an adult Elasmobranch, such as a dogfish.

34. The external and internal gills apparently form one continuous series of structures.
35. The lungs appear very early, and are used for a considerable period simultaneously with the gills.
36. The œsophagus becomes solid for a period commencing shortly before the mouth opens, and lasting some little time after the opening is effected.
37. The tadpole undergoes a distinct increase in bulk prior to the opening of the mouth. We have suggested that in the early stages nutriment is absorbed through the sucker.
38. There appear to be amongst Amphibians great differences between families, genera, species, and even individuals in the actual mode and relative times of development of the blood vessels of the mandibular, hyoidean, and branchial arches. Our present information is far too imperfect to enable us to determine which of these modes is to be regarded as typical for the group.

DESCRIPTION OF THE FIGURES ON PLATES XIII., XIV., AND XV.

Illustrating Professor MARSHALL and Mr. BLES' Paper on "The Development of the Blood Vessels in the Frog."

All the figures are from tadpoles of Rana temporaria.

Figs. 1, 2, and 3 are drawn from individual sections. All the remaining figures are diagrammatical, each being compounded from several sections. Parts are thus shown that really lie at very different levels, but in all cases the shapes and positions of the several organs are inserted from separate camera drawings of actual sections, and are represented as correctly as possible. Organs which are not immediately concerned with the blood vessels are in many cases omitted entirely.

In the majority of the figures the epiblast is represented black, and its division into sensory and nervous layers is not indicated. The efferent vessels of the visceral arches and the dorsal aorta are coloured red.

All the figures, with the exception of Fig. 15, are enlarged to the same extent, being magnified 45 diameters.

In all the figures showing the blood vessels from one side the brain and notochord are drawn from the median sagittal section of the series.

The small outline figures show the shapes of the tadpoles at the various stages, each of these figures being three times the natural size.

Alphabetical List of Reference Letters for all the Figures.

A	Dorsal aorta	A F'	Afferent vessel of first branchial arch: lacunar part
A B	Basilar artery		
A C	Internal carotid artery	A F. 2	Afferent vessel of second branchial arch: cardiac part
A E	Connection between the third branchial afferent and efferent vessels at base of gill	A F''	Afferent vessel of second branchial arch: lacunar part
A F.1	Afferent vessel of first branchial arch: cardiac part	A F. 3	Afferent vessel of third branchial arch: cardiac part

A F''' Afferent vessel of third branchial arch: lacunar part
A F.4   Afferent vessel of fourth branchial arch: cardiac part
A F'''' Afferent vessel of fourth branchial arch: lacunar part
A H   Afferent vessel of hyoid arch
A L   Lingual artery
A M   Anterior infundibular commissure
A M'   Posterior infundibular commissure
A O   Ophthalmic artery
A P   Pulmonary artery
A R   Thyroid branch of lingual artery
A S   Posterior palatine artery
A T   Anterior palatine artery
A U   Cutaneous artery
A Y   Pharyngeal artery
B F   Fore-brain
B H   Hind-brain
B J   Infundibulum
B M   Mid-brain
B P   Pineal body
B. 3   Choroid plexus of third ventricle
B. 4   Choroid plexus of fourth ventricle
C   Cœlom
C A   Pericardial cavity
C B.1   First branchial cleft
C B.2   Second ditto
C B.3   Third   ditto
C B.4   Fourth ditto
C G   Groove in floor of mouth, leading to first branchial cleft
C H   Hyomandibular cleft
C R   Carotid gland
E   Ear
E F.1   Efferent vessel of first branchial arch: aortic part
E F'   Efferent vessel of first branchial arch: lacunar part

E F.2   Efferent vessel of second branchial arch: aortic part
E F''   Efferent vessel of second branchial arch: lacunar part
E F.3   Efferent vessel of third branchial arch: aortic part
E F''' Efferent vessel of third branchial arch: lacunar part
E F.4   Efferent vessel of fourth branchial arch: aortic part
E F'''' Efferent vessel of fourth branchial arch: lacunar part
E H   Efferent vessel of hyoid arch: aortic part
E H'   Efferent vessel of hyoid arch: lacunar part
E M   Efferent vessel of mandibular arch
E R   Recessus labyrinthi, within skull
F. 1, 2, 3, 4   Filters of first, second, third, and fourth branchial arches
G   Glomerulus
G C   Capillary loop of external gill
G E   External gill
G L. 1, 2, 3, 4   Internal gills of first, second, third, and fourth branchial arches
H   Endothelial lining of heart
H A   Auricle
H B   Mesoblastic wall of heart
H L   Left auricle
H M   Truncus arteriosus: median part
H N   Ditto   lateral branch
H R   Right auricle
H S   Sinus venosus
H V   Ventricle
J   Jaw
K A   Auditory capsule
K B   Basihyal cartilage
K H   Cartilage of hyoid arch
K P   Parachordal

K.1	Cartilage of first branchial arch	Q	Intestine
		S	Sucker
K.4	Cartilage of fourth branchial arch	S T	Stomatodæum
		T	Thymus
L	Lung	V C	Posterior vena cava
L N	Lateral line	V D	Ductus Cuvieri, or Cuvierian vein
L P	Lip		
M O	Somatopleuric layer of mesoblast	V H	Hepatic vein
		V J	Inferior jugular vein
M P	Splanchnopleuric layer of mesoblast	V K	Venous sinus above sucker
		V M	Mandibular vein
N	Notochord	V P	Pulmonary vein
N O	Nephrostome	V R	Anterior cardinal vein
N P	Head kidney	V S	Posterior cardinal vein
O	Olfactory pit	V Y	Hyoidean vein
O C	Opercular cavity	W	Liver
O D	External aperture of opercular cavity	Y	Lymph space
		Z	Septum between stomatodæum and pharynx
O F	Opercular fold of hyoid arch		
O S	Optic stalk	VIII	Auditory nerve
P	Pharynx	X	Pneumogastric nerve
P B	Pituitary body		

Fig. 1. $3\frac{1}{4}$ mm. tadpole. Transverse section through the ear, showing the development of the pericardial cavity and of the endothelial lining of the heart: ($\times$ 45).

Fig. 2. $3\frac{1}{2}$ mm. tadpole. Median sagittal section of the anterior half of the tadpole, showing the relations of the heart to the pharynx and liver: ($\times$ 45).

Fig. 3. 4 mm. tadpole. Transverse section through the middle of the heart, showing the development of the heart and pericardial cavity: ($\times$ 45).

Fig. 4. $4\frac{1}{2}$ mm. tadpole. A diagrammatic figure of the anterior part of the tadpole, seen from the right side, showing the condition of the heart and blood vessels: ($\times$ 45).

Fig. 5. $4\frac{1}{2}$ mm. tadpole. A diagrammatic transverse section through the head, passing through the auditory vesicles, and seen from behind. The figure shows an abnormal condition, seen only in a single specimen of *Rana temporaria*, but apparently constant in *Rana esculenta*, in which the efferent vessel of the first branchial arch, which is shown on the left side of the figure, forms a direct communication between the truncus arteriosus and the aorta. On the right side the section passes through the first branchial gill pouch: (× 45).

Fig. 6. 5 mm. tadpole. A diagrammatic figure of the anterior part of the tadpole, seen from the right side, and showing the condition of the heart and blood vessels: (× 45).

Fig. 7. 6½ mm. tadpole. A diagrammatic figure of the anterior part of the tadpole, seen from the right side, and showing the condition of the heart and blood vessels: (× 45).

Fig. 8. 6½ mm. tadpole. A diagrammatic transverse section through the head, just behind the auditory vesicles, and seen from behind. The section passes through the first branchial arch on each side. On the right side, both the afferent and efferent vessels of this arch are shown; on the left side, the greater part of the afferent vessel has been removed in order to expose the efferent vessel more thoroughly: (× 45).

Fig. 9. 6½ mm. tadpole. A diagrammatic horizontal section through the anterior part of the tadpole. The general level of the section is that of the dorsal part of the pharynx just below its roof. On the right side, the section is taken at a level slightly ventral to that shown on the left side. On the left side, the whole course of the dorsal aorta and its branches are shown, together with the external gill of the second branchial arch, and the three nephrostomes of the head kidney. On the right side, the gill pouches are shown, together with the vessels of the several arches: (× 45).

Fig. 10. 6½ mm. tadpole. A diagrammatic horizontal section through the anterior part of the tadpole, taken at a level ventral to that of Fig. 9, and showing the heart and afferent branchial vessels as seen from above. On the right side the section is taken at a level slightly ventral to that shown on the left side: (× 45).

Fig. 11. 9 mm. tadpole. A diagrammatic figure of the anterior part of the tadpole, seen from the right side, and showing the condition of the heart and blood vessels: (× 45).

Fig. 12. 9 mm. tadpole. A diagrammatic horizontal section through the anterior part of the tadpole, showing the heart and the vessels, both afferent and efferent, of the first three branchial arches. On the left side, the section is taken at a more dorsal level than that of the right side, and the second and third branchial clefts are shown opening to the exterior: ($\times$ 45).

Fig. 13. 12 mm. tadpole. A diagrammatic figure of the anterior part of the tadpole, seen from the right side, and showing the condition of the heart and blood vessels: ($\times$ 45).

Fig. 14. 12 mm. tadpole. A diagrammatic horizontal section through the anterior part of the tadpole, showing the heart and the afferent and efferent branchial vessels in the floor of the pharynx, seen from above: ($\times$ 45).

Fig. 15. 23 mm. tadpole. A diagrammatic transverse section through the head, passing through the auditory organs, showing the heart and branchial vessels in section. On the left side, the section is taken at a level rather further back than that of the right side of the figure: ($\times$ 22).

Plate XIII

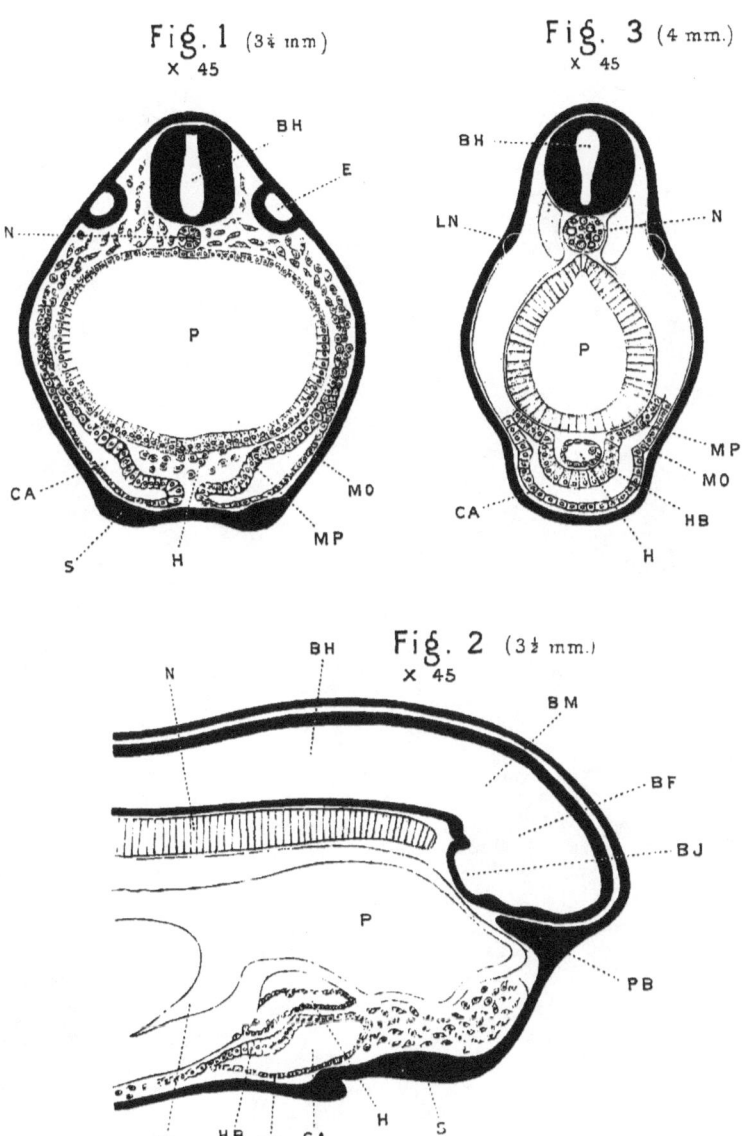

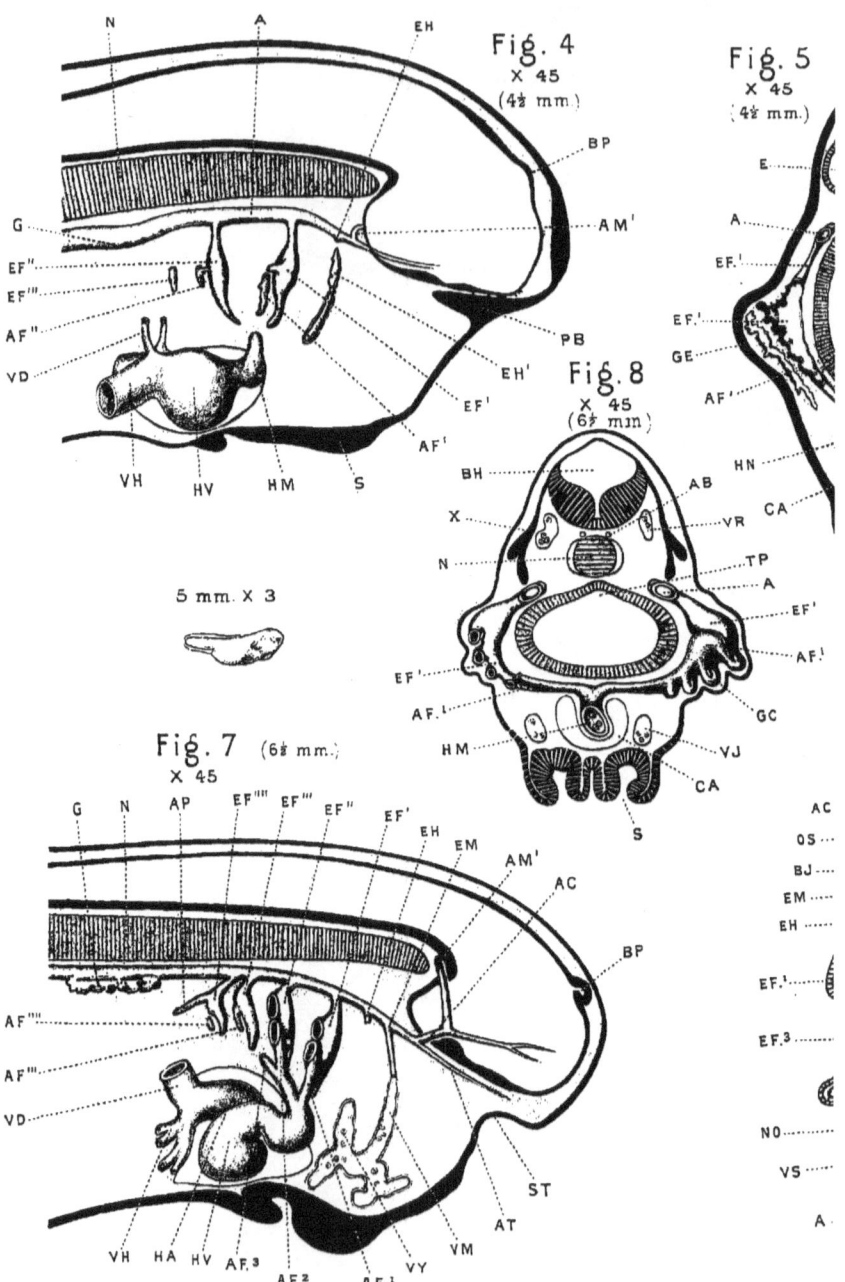

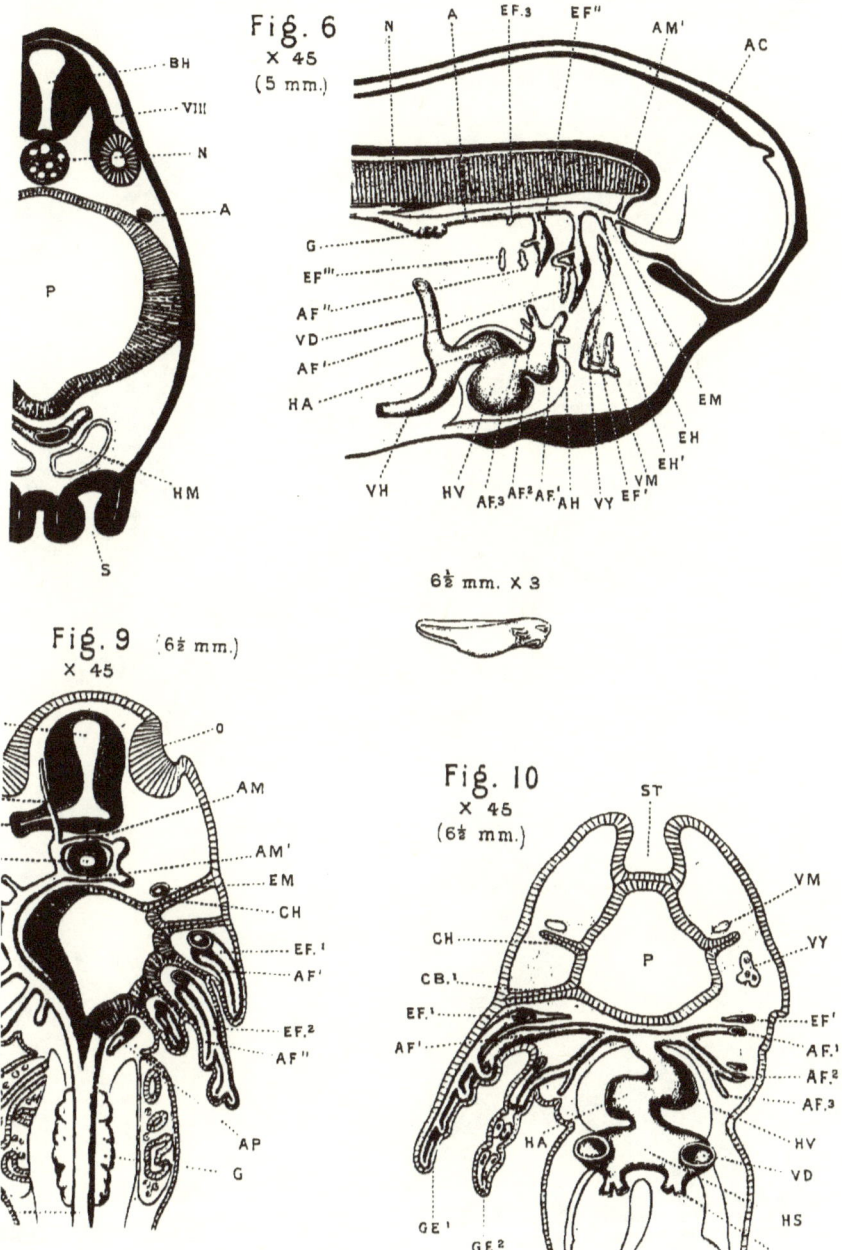

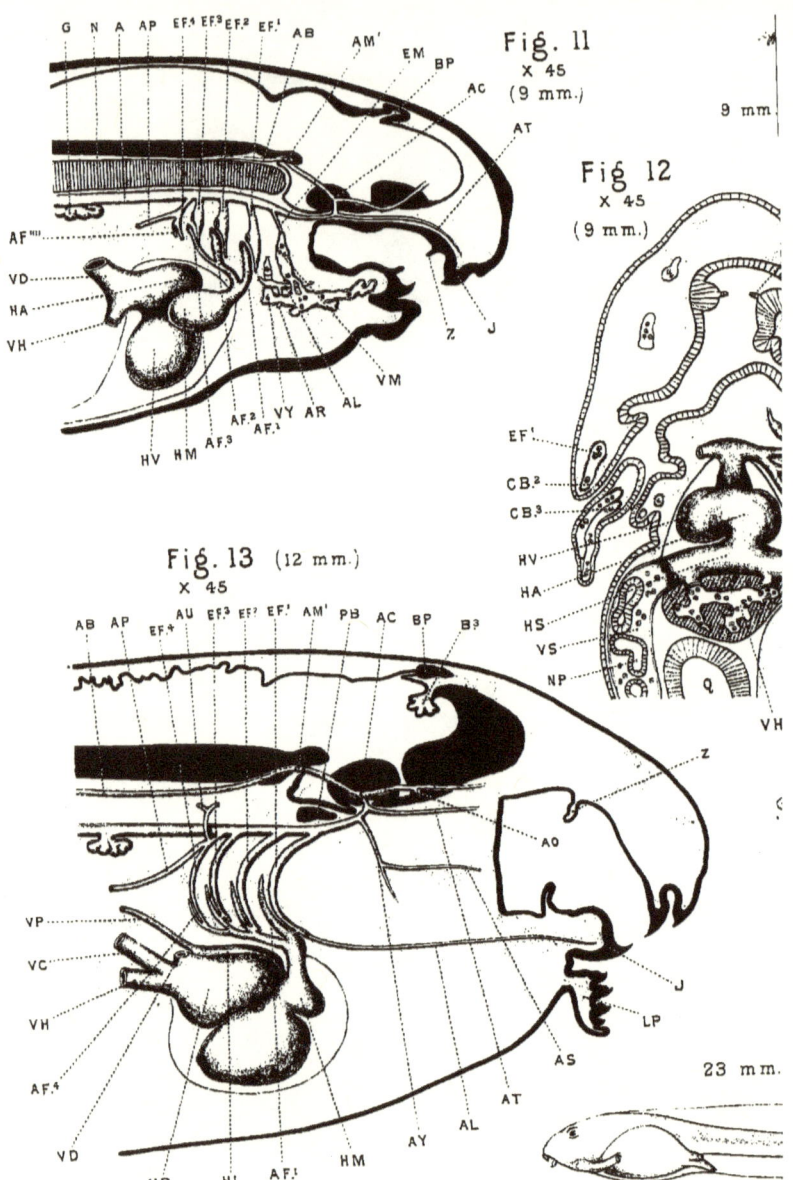

Plate XV.

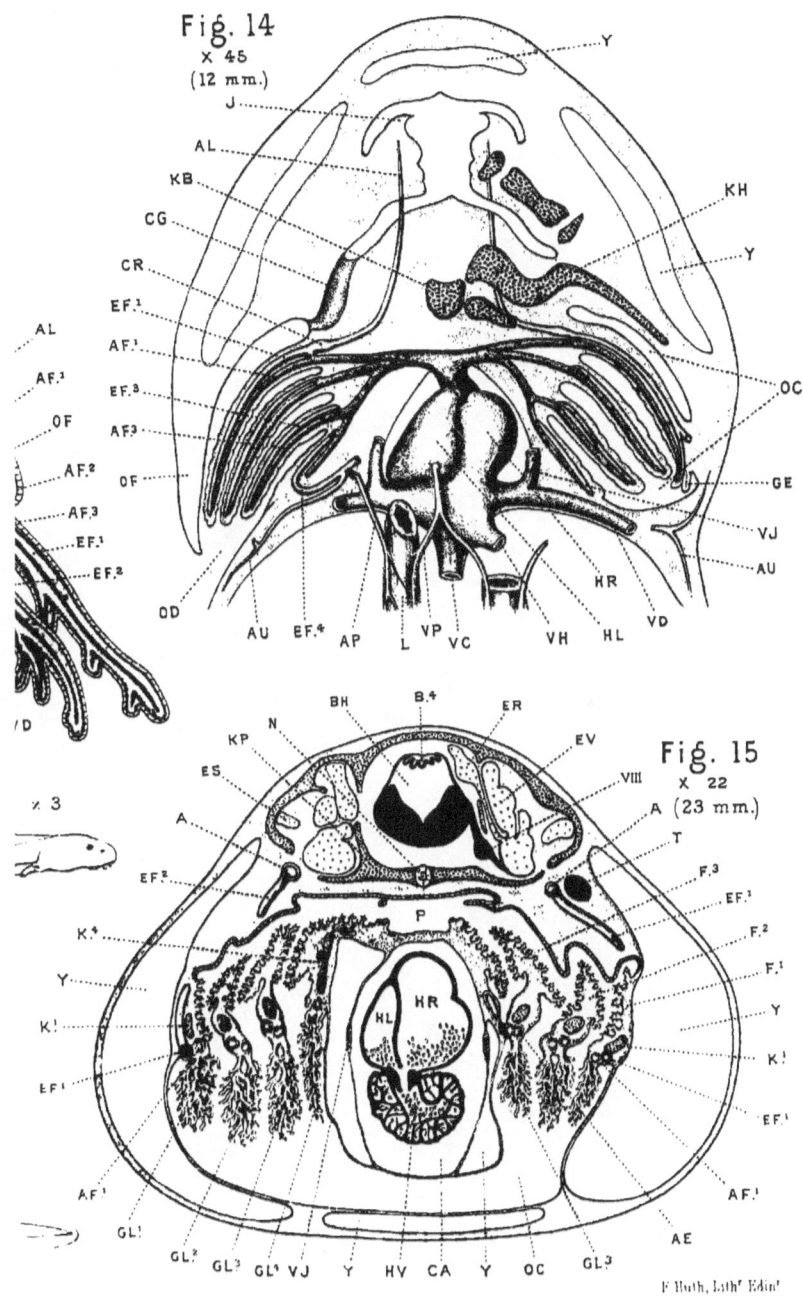

www.ingramcontent.com/pod-product-compliance
Lightning Source LLC
Chambersburg PA
CBHW030013240426
43672CB00007B/927